LONDON MATHEMATICAL SOCIETY LEC

Managing Editor: Professor J.W.S. Cassels, Departmen   :s,
University of Cambridge, 16 Mill Lane, Cambridge CB

The titles below are available from booksellers, or, in ca   s.

London Mathematical Society Lecture Note Series. 203

Hochschild Cohomology of von Neumann Algebras

Allan M. Sinclair
University of Edinburgh

Roger R. Smith
Texas A & M University

CAMBRIDGE
UNIVERSITY PRESS

Published by the Press Syndicate of the University of Cambridge
The Pitt Building, Trumpington Street, Cambridge CB2 1RP
40 West 20th Street, New York, NY 10011-4211, USA
10 Stamford Road, Oakleigh, Melbourne 3166, Australia

First published 1995

Printed in Great Britain at the University Press, Cambridge

Library of Congress cataloging in publication data
Sinclair, Allan M.
Hochschild cohomology of von Neumann algebras / by Allan M.Sinclair
and Roger R. Smith.
    p.   cm. - (London Mathematical Society lecture note series; 203)
Includes bibliographical references and index
ISBN 0 521 47880 4 (pbk.)
1. Homology theory. 2. Von Neumann algebras. I. Smith, Roger R.
II. Title. III. Series.
QA612.3.S56 1995
514'.23-dc20   94-38845 CIP

British Library cataloguing in publication data available

ISBN   0 521 47880 4 paperback

Contents

Preface

The authors wish to thank all their friends and colleagues who sent them reprints and preprints, and explained parts of the theory to them. Barry Johnson, Dick Kadison and John Ringrose laid the foundation of the subject and wrote about it clearly. We are indebted to them, and particularly to John Ringrose for the two survey articles [Ri3, Ri6] which made the subject accessible and are still essential reading. Many of the beautiful and deep ideas in Chapters 1, 4 and 6 are due to Erik Christensen on his own or in collaboration with one or both authors. Without his mathematical insight, perseverance and encouragement this book would not have been written. It is a great pleasure to acknowledge our deep debt to him.

The patience and understanding of our wives Patricia and Virginia have been invaluable, and we have received much support and advice from Roger Astley and David Tranah of Cambridge University Press. Our special thanks go to Robin Campbell who typed the entire manuscript with great expertise.

While writing this book, both authors were supported by a NATO collaborative research grant, and the second author by a grant from the National Science Foundation. We take this opportunity to record our sincere gratitude to both agencies.

0

Introduction

0.1 General Introduction

The theory of multilinear maps on a von Neumann algebra is developed in these notes and applied to the continuous Hochschild cohomology of von Neumann algebras. The methods used are those of von Neumann algebras and complete boundedness rather than of homological algebra, and only elementary cohomlogical techniques are employed in the proofs. We have chosen to base our presentation on the problem of whether the continuous cohomology groups $H^n(M, M)$ of a von Neumann algebra M over itself are zero for all n. This, and closely related questions, has stimulated much of the recent development of the theory of completely bounded maps, and so we have adopted an approach which has wider applications beyond cohomology theory. The results in these notes have been proved in full generality, provided that they do not stray too far from the central topic of dual normal modules over von Neumann algebras.

There are two main reasons for investigating the Hochschild cohomology groups of operator algebras. When they are non-zero they provide invariants which can distinguish classes of algebras; when they are zero they lead to positive results on the stability of algebraic structures and on the space of bounded derivations on an operator algebra. Elliott's classification of separable AF-C^*-algebras by K-theory is an example of the use of a homological invariant [El1]. Connes has characterized the injective von Neumann algebras by the vanishing of their cohomology over all dual normal modules [Co1], while a further example is the equivalence of amenability and nuclearity for C^*-algebras, established by Haagerup [Ha1]. In each of these latter results cohomological conditions are closely related to metric properties of von Neumann algebras and C^*-algebras. This theme will recur subsequently in these notes when geometrical ideas and methods are used to deduce information about cohomology groups.

Before presenting an outline of the contents of these notes, we introduce the Hochschild complex and give a brief history of the subject. The reader will find more detailed historical remarks at the end of each chapter.

0.2 The Hochschild Complex

Let M be a von Neumann algebra and let X be a Banach M-bimodule. By this we mean that there is a module action of M on both the left and right of X satisfying

$$\|mx\| \leq \|m\| \, \|x\| \quad \text{and} \quad \|xm\| \leq \|x\| \, \|m\|$$

for all $m \in \mathcal{M}, x \in \mathcal{X}$. We assume throughout that

$$1x = x1 = x$$

for all $x \in \mathcal{X}$, where 1 is the identity of $\mathcal{M}$. We will be concerned with two main examples. The bimodule may be $\mathcal{M}$ itself, or $B(H)$ if $\mathcal{M}$ is represented on a Hilbert space H; in both cases the module action is the standard multiplication of operators. The space of $\mathcal{X}$-valued continuous n-linear maps on the n-fold Cartesian product $\mathcal{M}^n = \mathcal{M} \times \cdots \times \mathcal{M}$ is denoted by $\mathcal{L}^n(\mathcal{M}, \mathcal{X})$ for $n \geq 1$, while $\mathcal{L}^0(\mathcal{M}, \mathcal{X})$ is defined to be $\mathcal{X}$.

The coboundary operator $\partial \colon \mathcal{L}^n(\mathcal{M}, \mathcal{X}) \to \mathcal{L}^{n+1}(\mathcal{M}, \mathcal{X})$ is defined as follows. For $n = 0$,

$$(\partial x)(a) = ax - xa \qquad (x \in \mathcal{X}, a \in \mathcal{M}),$$

while for $n \geq 1$,

$$(\partial \phi)(a_1, \ldots, a_{n+1}) = a_1 \phi(a_2, \ldots, a_{n+1})$$
$$+ \sum_{j=1}^{n} (-1)^j \phi(a_1, \ldots, a_{j-1}, a_j a_{j+1}, a_{j+2}, \ldots, a_{n+1})$$
$$+ (-1)^{n+1} \phi(a_1, \ldots, a_n) a_{n+1}$$

where $\phi \in \mathcal{L}^n(\mathcal{M}, \mathcal{X})$ and $a_1, \ldots, a_{n+1} \in \mathcal{M}$. It is easy to check that $\partial^2 \colon \mathcal{L}^n(\mathcal{M}, \mathcal{X}) \to \mathcal{L}^{n+2}(\mathcal{M}, \mathcal{X})$ is always zero. For example, if $n = 0$ then

$$(\partial^2 x)(a_1, a_2) = a_1(a_2 x - xa_2) - (a_1 a_2 x - xa_1 a_2) + (a_1 x - xa_1)a_2$$
$$= 0.$$

For the general case, let $\psi_0, \ldots, \psi_{n+1}$ denote the $(n+1)$-linear maps in the definition of $\partial \phi$, so that

$$\partial \phi = \psi_0 + \sum_{j=1}^{n} \psi_j + \psi_{n+1}.$$

There are six types of terms in the sum for $\partial^2 \phi$. We show that they occur in pairs with opposite signs, proving that $\partial^2 \phi = 0$:

(i) $a_1 a_2 \phi(a_3, \ldots, a_{n+2})$ occurs twice in $\partial \psi_0$ as

$$a_1 \psi_0(a_2, \ldots, a_{n+2}) - \psi_0(a_1 a_2, a_3, \ldots, a_{n+2}),$$

(ii) $a_1 \phi(a_2, \ldots, a_{j-1}, a_j a_{j+1}, a_{j+2}, \ldots, a_{n+2})$ occurs with coefficients $(-1)^j$ and $(-1)^{j-1}$ respectively in $\partial \psi_0$ and $\partial \psi_{j-1}$,

(iii) $\phi(a_1, \ldots, a_{j-1}, a_j a_{j+1}, a_{j+2}, \ldots, a_{r-1}, a_r a_{r+1}, a_{r+2}, \ldots, a_{n+2})$ occurs with coefficients $(-1)^{j+r-1}$ and $(-1)^{r+j}$ respectively in $\partial\psi_j$ and $\partial\psi_r$,

(iv) $\phi(a_1, \ldots, a_{j-1}, a_j a_{j+1}, a_{j+2}, a_{j+3}, \ldots, a_{n+2})$ occurs with coefficients $(-1)^{2j}$ and $(-1)^{2j+1}$ respectively in $\partial\psi_j$ and $\partial\psi_{j+1}$,

(v) $\phi(a_1, \ldots, a_{n-1}, a_n a_{n+1}) a_{n+2}$ occurs with coefficients $(-1)^{2n+2}$ and $(-1)^{2n+1}$ respectively in $\partial\psi_n$ and $\partial\psi_{n+1}$,

(vi) $\phi(a_1, \ldots, a_n) a_{n+1} a_{n+2}$ occurs twice in $\partial\psi_{n+1}$ as

$$(-1)^{2n+2}\psi_{n+1}(a_1, \ldots, a_n, a_{n+1}a_{n+2}) + (-1)^{2n+3}\psi_{n+1}(a_1, a_2, \ldots, a_{n+1})a_{n+2}.$$

The reader may find it helpful to write out the case $n = 1$ explicitly. The (continuous) Hochschild complex for $\mathcal{M}$ acting on $\mathcal{X}$ is

$$\mathcal{L}^0(\mathcal{M}, \mathcal{X}) \xrightarrow{\partial} \mathcal{L}^1(\mathcal{M}, \mathcal{X}) \xrightarrow{\partial} \mathcal{L}^2(\mathcal{M}, \mathcal{X}) \xrightarrow{\partial} \cdots$$

where, from above,

$$\mathrm{Im}(\partial\colon \mathcal{L}^{n-1}(\mathcal{M}, \mathcal{X}) \to \mathcal{L}^n(\mathcal{M}, \mathcal{X})) \subseteq \mathrm{Ker}(\partial\colon \mathcal{L}^n(\mathcal{M}, \mathcal{X}) \to \mathcal{L}^{n+1}(\mathcal{M}, \mathcal{X})).$$

The nth Hochschild cohomology group $H^n(\mathcal{M}, \mathcal{X})$ is then defined to be the quotient vector space (which we regard as an additive abelian group)

$$\frac{\mathrm{Ker}(\partial\colon \mathcal{L}^n(\mathcal{M}, \mathcal{X}) \to \mathcal{L}^{n+1}(\mathcal{M}, \mathcal{X}))}{\mathrm{Im}(\partial\colon \mathcal{L}^{n-1}(\mathcal{M}, \mathcal{X}) \to \mathcal{L}^n(\mathcal{M}, \mathcal{X}))}$$

for $n \geq 1$. Elements of the kernel are called cocycles and elements of the image are called coboundaries.

These groups contain a considerable amount of information about the von Neumann algebra $\mathcal{M}$. Consider the case $n = 1$. Then $\mathrm{Ker}\,\partial$ is the space of maps $\phi\colon \mathcal{M} \to \mathcal{X}$ satisfying

$$a_1\phi(a_2) - \phi(a_1 a_2) + \phi(a_1)a_2 = 0 \qquad (a_1, a_2 \in \mathcal{M})$$

and this is precisely the space of bounded derivations. On the other hand, $\mathrm{Im}\,\partial$ is the space of inner derivations of $\mathcal{M}$ into $\mathcal{X}$, and so $H^1(\mathcal{M}, \mathcal{X})$ measures how close derivations are to being inner. The higher-order groups do not have such a familiar interpretation, but nevertheless we will see in Chapter 7 that they play an important role in the structure theory of von Neumann algebras.

In the theory developed here, it is important to pass from the complex $(\mathcal{L}^n(\mathcal{M}, \mathcal{X}), \partial)$ to more specialized subcomplexes. A particularly important case is $(\mathcal{L}^n_w(\mathcal{M}, \mathcal{X}), \partial)$ where the subscript "w" indicates ultraweak to weak* continuity of the n-linear maps. This gives rise to normal cohomology groups

$H_w^n(\mathcal{M}, \mathcal{X})$, which may be easier to compute. Indeed, a major theme in these notes is to calculate $H^n(\mathcal{M}, \mathcal{X})$ indirectly by first proving that

$$H^n(\mathcal{M}, \mathcal{X}) \cong H_*^n(\mathcal{M}, \mathcal{X})$$

for the cohomology groups of a suitable subcomplex $(\mathcal{L}_*^n(\mathcal{M}, \mathcal{X}), \partial)$ and then determining $H_*^n(\mathcal{M}, \mathcal{X})$. Another important example is the subcomplex obtained by considering n-linear maps which are modular with respect to suitable subalgebras. Such maps arise by averaging over amenable unitary groups of operators, and we will devote considerable time to this crucial topic.

0.3 History

In [Kap] Kaplansky asked various questions about the properties of derivations on C^*-algebras and von Neumann algebras, which may be interpreted as inquiring if certain cohomology groups are equal, or equal to zero. He then established some results for type I algebras. After preliminary work by Kadison [Ka2], Sakai [S1] showed that all the derivations on a von Neumann algebra $\mathcal{M}$ are inner, which is equivalent to showing that the first continuous cohomology group $H^1(\mathcal{M}, \mathcal{M})$ is zero. Of course all of this had been preceded by Hochschild's research on the homology and cohomology of rings and algebras [Ho1, Ho2, Ho3]. Here we limit the discussion to von Neumann algebras and their modules and maps. From 1968 to 1972 Johnson, Kadison and Ringrose [KR2, KR3, JKR, J5] proved a number of technical results from which they deduced the equality of certain cohomology groups associated with different classes of continuous maps from the algebra into suitable Banach modules. In particular they showed that a hyperfinite von Neumann algebra $\mathcal{M}$ is amenable as a von Neumann algebra, in the sense that the first cohomology into all dual normal modules is zero. From this it follows that $H^n(\mathcal{M}, \mathcal{M}) = 0$ for all $n \geq 1$ when $\mathcal{M}$ is hyperfinite [JKR]. The last two authors conjectured that these groups are zero for all von Neumann algebras $\mathcal{M}$. This conjecture is still open at the time of writing (1994) and it will be discussed more fully subsequently. Their techniques are still important today, and will be presented in Chapter 3 and Section 5.2.

The current theory has taken shape in the last ten years, but there are several other results prior to the early 1980s which should be mentioned. Connes' characterization of injective von Neumann algebras as those which are amenable [Co1] had already been hinted at in his proof of the equivalence of injectivity and hyperfiniteness [Co2]. There were also some investigations of cohomology with the algebra of compact operators as the module [JP, PopR]. In these notes we will only consider dual normal modules, and we will regard more general modules as being beyond our scope. Developments in the theory of completely bounded maps [ChS1, PS], which were partly motivated by cohomology, led to further progress in calculating the cohomology with the algebra $B(H)$ of bounded operators on a Hilbert space as the module [ChES]. These authors introduced completely bounded cohomology groups, showed that they were zero, and in many cases were able to prove that they were equal to the continuous cohomology groups (see Chapter 6). These results established complete boundedness as fundamental to cohomology theory. It enters the picture when a von Neumann algebra is stable under tensoring with $B(H)$ (the type I, II_∞ and III cases) or with the hyperfinite type II_1 factor $\mathcal{R}$ (many examples in the type II_1 case). Perhaps the most interesting and least understood situation is that of a type II_1 factor $\mathcal{M}$ which is not isomorphic to $\mathcal{M}\overline{\otimes}\mathcal{R}$.

The Hochschild cohomology of Banach algebras developed partly in advance of operator algebra theory and partly behind it. We refer the reader to the memoir by Johnson [J3] and the book by Helemskii [He] for the details. The amenable aspects of von Neumann algebras are discussed in these notes only to the extent that they motivate the calculation of $H^n(\mathcal{M}, \mathcal{V})$ for dual normal modules $\mathcal{V}$ over $\mathcal{M}$. The books of Paterson [P1] and Pier [Pie] or the unpublished notes by Thorpe [Th] contain full discussions of amenability. We have omitted the purely algebraic side of the theory in which no assumptions of continuity are made for the cocycles and coboundaries. For this the reader should consult the work of Wodzicki [Wo].

0.4 Averaging

The basic idea behind the calculation of cohomology groups is to take an average of a suitable type. This allows us to reduce a general cocycle to one which has desirable algebraic and continuity properties. For a von Neumann algebra this generally means modularity with respect to some hyperfinite subalgebra and normality with respect to each variable.

To motivate the averaging process, consider a von Neumann algebra $\mathcal{M}$, a subalgebra $\mathcal{A}$ generated by a norm compact unitary subgroup $\mathcal{U}$ with normalized Haar measure μ, and a derivation (1-cocycle) δ of $\mathcal{M}$ into a Banach module $\mathcal{X}$. If $u, v \in \mathcal{U}$, then

$$\delta(uv) = u\delta(v) + \delta(u)v$$

so

$$\delta(v) = u^*\delta(uv) - u^*\delta(u)v$$
$$= v(uv)^*\delta(uv) - u^*\delta(u)v.$$

Integrating with respect to Haar measure and using invariance leads to

$$\delta(v) = vx_0 - x_0v$$

where $x_0 = \int_{\mathcal{U}} u^* \delta(u) d\mu(u)$. This integral is a norm limit of finite approximating sums so defines an element of $\mathcal{X}$. After subtraction of the inner derivation defined by x_0 we obtain a new derivation $\tilde{\delta}$ which vanishes on $\mathcal{A}$. The defining equation for a derivation then gives modularity with respect to $\mathcal{A}$:

$$\tilde{\delta}(am) = a\tilde{\delta}(m) + \tilde{\delta}(a)m = a\tilde{\delta}(m),$$
$$\tilde{\delta}(ma) = m\tilde{\delta}(a) + \tilde{\delta}(m)a = \tilde{\delta}(m)a,$$

for $m \in \mathcal{M}$, $a \in \mathcal{A}$. The classical algebraic case of a finite group is a special case; Haar measure becomes a finite averaging process over the group elements. For general von Neumann algebras we will need to move beyond Haar measure and so $\int_{\mathcal{U}} \dots d\mu(u)$ may denote an average by an invariant mean when $\mathcal{U}$ is an amenable unitary group. In this situation the integral will converge in the weak*-topology, forcing us to require $\mathcal{X}$ to be a dual space. Moreover, to pass from $\mathcal{U}$ to the von Neumann algebra it generates requires ultraweak limits and compatibility between the ultraweak topology of $\mathcal{M}$ and the weak*-topology of the module $\mathcal{X}$. For this reason we restrict attention to dual normal modules: $\mathcal{X}$ is a dual space and the maps

$$m \to mx \quad \text{and} \quad m \to xm$$

are both ultraweak-weak* continuous from $\mathcal{M}$ into $\mathcal{X}$ for each fixed element $x \in \mathcal{X}$.

Averages are usually taken over unitary groups, but we will also discuss averages based on projections in Sections 1.7 and 4.2.

0.5 Contents

We give only a brief review of the contents, since each chapter begins with an introduction and ends with notes and remarks.

Chapter 1 is devoted to the theory of completely positive and completely bounded maps, and of the Haagerup tensor product. These are crucial tools from Chapter 4 onwards. The representation theorem for completely bounded multilinear maps is proved in detail using the Haagerup tensor product. We follow the operator space approach of [PS] rather than the original one in [ChS1]. The development of complete boundedness occupies the first six sections, and we conclude the chapter by showing the existence of a useful projection from the space of completely bounded maps into a von Neumann algebra $\mathcal{M}$ onto the space of those maps which are modular with respect to $\mathcal{M}$. We subsequently apply this result to show that derivations on von Neumann algebras are inner (Theorem 2.5.1), and that the completely bounded cohomology of a von Neumann algebra over itself is zero (Theorem 4.3.1).

Chapter 2 gives a short account of the theory of derivations on von Neumann algebras into themselves. This is just intended as an introduction to cohomology since there are good detailed accounts of this elsewhere [Di2, S3]. We restrict ourselves to bounded derivations and go only as far as the Kadison–Sakai theorem [Ka2, S1] that derivations on von Neumann algebras are inner. Many readers will be familiar with this. However, our approach is not the standard one, and the techniques in this chapter will occur subsequently in more complicated form.

The third chapter covers the material developed by Johnson, Kadison and Ringrose in the period 1968–1972. Our account has been strongly influenced by two survey articles by Ringrose [Ri3, Ri6]. The notation is different but the organization is the same. The fundamental idea is to average maps over the amenable unitary group of a suitable C^*-subalgebra of the given C^*-algebra or von Neumann algebra. The averaged map is then used in elementary, but tricky, calculations to show that cohomology groups associated with module maps over the averaged algebra are equal to the unaveraged groups. Alternatively, we could have chosen the more axiomatic and homological approach due to Craw [Cr1, Cr2]. This works well in the continuous and normal cases, but there are major technical problems in the completely bounded theory. The continuous setting requires the duality between the projective tensor product and the space of continuous bilinear forms. In the completely bounded case, all the Banach spaces must be replaced by operator spaces while retaining the correct dualities and relations. This is now possible, due to the theory of operator spaces and their tensor products as recently developed by Blecher and Paulsen [BP] and Effros and Ruan [ER1–ER5] (but which was not available to Craw). We decided that the axiomatic approach would be less intuitive and would require more operator space theory than we thought appropriate.

In Chapter 4 we show that the completely bounded cohomology of a von Neumann algebra $\mathcal{M}$ over $B(H)$ is always zero. This depends on writing cocycles as products of commutators and using a projection as an average. The corresponding result when $\mathcal{M}$ is the module is deduced from this, using the projection of Section 1.7. The representation of a cocycle as a product of commutators is only available in the completely bounded case. For this reason, all recent progress in the continuous setting depends on reduction to this special situation.

Chapter 5 is concerned with hyperfinite subalgebras and maximal abelian self-adjoint subalgebras (masas) in type II_1 von Neumann algebras; this is the technical foundation for the succeeding chapter. Section 5.2 contains the important result that a continuous multilinear map on a von Neumann algebra $\mathcal{M}$ to itself which is modular over the centre can be continuously extended to the C^*-algebra generated by $\mathcal{M}$ and a masa in the commutant

$\mathcal{M}'$. The point is that the ultraweak closure of this C^*-algebra is a type I von Neumann algebra, which is much easier to handle than $\mathcal{M}$ itself.

The following section is devoted to three results of Popa on subalgebras of type II_1 factors. The first two provide a hyperfinite subalgebra $\mathcal{N}$ whose relative commutant $\mathcal{N}' \cap \mathcal{M}$ consists of scalar multiples of the identity. These results provide the subalgebras over which we average in Sections 5.5 and 6.3. The third shows how to obtain a masa in $B(L^2(\mathcal{M}))$ from a Cartan subalgebra in $\mathcal{M}$. In Section 5.4 we introduce the Haagerup–Pisier non-commutative Grothendieck inequality. The normal form of this is used to show that certain bimodule maps lift to the Haagerup tensor product. Combined with results from Section 1.6, this is an important step towards proving that certain cocycles are completely bounded in Sections 6.3 and 6.4.

The main calculations of cohomology groups are presented in Chapter 6. Type I_∞, II_∞, and III von Neumann algebras are isomorphic to their tensor products with $B(H)$, while many type II_1 von Neumann algebras are isomorphic to their tensor product with $\mathcal{R}$. In both cases the extra tensor factor allows us to replace a general cocycle by a completely bounded one, after which the theory of Chapter 4 can be applied to prove that $H^n(\mathcal{M}, \mathcal{M}) = 0$ for these algebras. The type II_1 case is still open but we use complete boundedness to determine $H^2(\mathcal{M}, \mathcal{M})$ and $H^3(\mathcal{M}, \mathcal{M})$ when there is a Cartan subalgebra (Section 6.3) and $H^2(\mathcal{M}, \mathcal{M})$ for property Γ factors (Sections 6.4). We conclude the chapter with a general result which shows that it is sufficient to consider the conjecture "$H^n(\mathcal{M}, \mathcal{M}) = 0$" only for separably acting von Neumann algebras.

Chapter 7 presents that part of the structure theory of von Neumann algebras which depends on cohomology, and is drawn from [J6, RaT]. The main result is local stability of the product in a von Neumann algebra whenever the second and third cohomology groups vanish. The appendix (Chapter 8) contains a short discussion of bounded group cohomology of a discrete group G, and its relationship to the Hochschild cohomology of the Banach algebra $\ell^1(G)$. At this time there is no relation to the cohomology of group C^*-algebras and von Neumann algebras, but we have included this chapter in the hope that a connection may become apparent in the future.

0.6 Background

We have attempted to make these notes as self-contained as possible, and so we have assumed that the reader has had a first course in operator algebras but little beyond that. Knowledge of the basic theory as set out in the book of Kadison and Ringrose [KR4] (or the standard texts of Dixmier [Di2], Sakai [S2] or Takesaki [T]) would be ideal; however much less is required. The reader who knows the GNS representation, type theory of von Neumann algebras, the double commutant theorem, the Kaplansky density theorem, and the various topologies on a von Neumann algebra is ready to begin. We review briefly a few topics which will recur frequently but would be considered beyond basic knowledge.

A von Neumann algebra $\mathcal{M} \subseteq B(H)$ is said to be *hyperfinite* if there is an increasing family of finite dimensional $*$-subalgebras $\mathcal{M}_\lambda$ whose union is ultraweakly dense. If there is a norm one projection of $B(H)$ onto $\mathcal{M}$ then $\mathcal{M}$ is called *injective*. Connes [Co2] proved that these conditions are equivalent and so we use them interchangeably in the text. We will generally say "hyperfinite" if we want to use the finite dimensional subalgebras, and switch to "injective" when we need the norm one projection.

In Chapter 6 we will require some deeper theory for type II_1 factors. These are characterized by the existence of a tracial state tr satisfying

$$tr(xy) = tr(yx)$$

for all elements x, y of the factor $\mathcal{M}$. The GNS construction for tr yields a Hilbert space $L^2(\mathcal{M}, tr)$ (or just $L^2(\mathcal{M})$) on which $\mathcal{M}$ acts by left multiplication. Since

$$tr(xx^*) = tr(x^*x),$$

the map $x \to x^*$ extends to a conjugate linear isometry J on $L^2(\mathcal{M})$ and $J\mathcal{M}J$ is the commutant $\mathcal{M}'$. This representation is the *standard form* for $\mathcal{M}$. The Hilbert space norm is denoted $\|\cdot\|_2$ and is defined by

$$\|x\|_2 = (tr(x^*x))^{1/2}.$$

If $\mathcal{N}$ is a subfactor of $\mathcal{M}$ then there always exists a unique trace-preserving conditional expectation $E\colon \mathcal{M} \to \mathcal{N}$, that is, a norm one projection such that

$$tr_{\mathcal{N}}(Ex) = tr_{\mathcal{M}}(x)$$

for all $x \in \mathcal{M}$. These facts about type II_1 von Neumann algebras may all be found in [KR4].

Cohomology theory is really the study of multilinear maps of a von Neumann algebra $\mathcal{M}$ into a module $\mathcal{X}$, so we review the classes of maps which will be important. The space of continuous multilinear maps is denoted by

$$\mathcal{L}^n(\mathcal{M}, \mathcal{X}).$$

If we restrict to maps which are modular with respect to a subalgebra $\mathcal{A}$ in the sense that

$$a\phi(x_1, \ldots, x_n) = \phi(ax_1, \ldots, x_n),$$

$$\phi(x_1, \ldots, x_j a, x_{j+1}, \ldots, x_n) = \phi(x_1, \ldots, x_j, ax_{j+1}, \ldots, x_n),$$

$$\phi(x_1, \ldots, x_n a) = \phi(x_1, \ldots, x_n)a$$

for $a \in \mathcal{A}$, $x_j \in \mathcal{M}$, then we adopt the notation

$$\mathcal{L}^n(\mathcal{M}, \mathcal{X}: /\mathcal{A}).$$

We append the subscripts "w" and "cb" to indicate ultraweak continuity in each variable and complete boundedness respectively. For example, $\mathcal{L}^n_{wcb}(\mathcal{M}, \mathcal{X}: /\mathcal{A})$ denotes the space of ultraweak-weak* continuous (equivalently normal) completely bounded n-linear maps of $\mathcal{M}$ into $\mathcal{X}$ which are modular with respect to $\mathcal{A}$. The cohomology groups $H^n_{wcb}(\mathcal{M}, \mathcal{X}: /\mathcal{A})$ are formed from cocycles and coboundaries from $\mathcal{L}^n_{wcb}(\mathcal{M}, \mathcal{X}: /\mathcal{A})$, with obvious similar interpretations of $H^n_{cb}(\mathcal{M}, \mathcal{X})$, $H^n_w(\mathcal{M}, \mathcal{X})$ and $H^n_{wcb}(\mathcal{M}, \mathcal{X})$. A general list of notation is included at the end of these notes.

1

Completely Bounded Operators

1.1 Introduction

This chapter contains much of the background material which is necessary for the study of the cohomology theory of von Neumann algebras. In Sections 1.2 and 1.3 we introduce the basic concepts: operator systems, operator spaces, completely positive maps and completely bounded maps. The two fundamental results in the subject are the Stinespring representation theorem (Theorem 1.2.1) and Arveson's Hahn–Banach theorem for completely positive maps (Theorem 1.2.3). We then discuss matrix ordered spaces, and obtain an important abstract characterization of operator systems (Theorem 1.2.7). With these results established, the representation of a completely bounded map as $V^*\pi W$ (Theorem 1.3.1) is easily obtained.

The fourth section is devoted to the Haagerup tensor product of operator spaces, in preparation for the succeeding section where complete boundedness is introduced for multilinear maps. The point is that multilinear maps can be viewed as linear maps on tensor products. The Haagerup tensor product norm is the correct one for compatibility with the completely bounded norm, and this allows us to prove multilinear results by appealing to the linear theorems of the second and third sections. The most important theorems here are 1.5.6 and 1.5.8. The first gives a general representation theorem for multilinear completely bounded maps on operator spaces, while the second describes an improved version for completely bounded maps on von Neumann algebras which are separately normal in each variable.

There are no known representation theorems for general bounded linear or multilinear maps and so it is useful to have available results which deduce complete boundedness from extra algebraic conditions. In Section 1.6 we present two such sets of hypotheses. The first (Theorem 1.6.1) asserts that a bounded linear map which is modular with respect to a C^*-algebra possessing a cyclic vector is completely bounded. The second (Theorem 1.6.2) is a variation on this theme for bilinear maps.

The final section of the chapter is concerned with this situation: $\mathcal{M} \subseteq \mathcal{A} \subseteq B(H)$, where $\mathcal{M}$ is a von Neumann algebra, $\mathcal{A}$ is a C^*-algebra, and we study the space of completely bounded maps of $\mathcal{A}$ into $\mathcal{M}$. The main result states that there is a contractive projection of this space onto the subspace of completely bounded right $\mathcal{M}$-module maps into $\mathcal{M}$. We will need this for the discussion of derivations and cohomology into the algebra in subsequent chapters.

1.2 Completely Positive Maps

If $\mathcal{A}$ is a C^*-algebra faithfully represented on a Hilbert space H then we may regard an $n \times n$ matrix with entries from $\mathcal{A}$ as an operator on the n-fold direct sum of copies of H. The operator norms on $B(H \oplus \cdots \oplus H)$ thus induce C^*-algebra norms on the matrix algebras $\mathsf{M}_n(\mathcal{A})$ over $\mathcal{A}$, and they are independent of the particular faithful representation of $\mathcal{A}$ used to define them. Any linear map $\phi \colon \mathcal{A} \to \mathcal{B}$ between C^*-algebras gives a family $\{\phi_n \colon \mathsf{M}_n(\mathcal{A}) \to \mathsf{M}_n(\mathcal{B})\}_{n=1}^{\infty}$ defined by $\phi_n(a_{ij}) = (\phi(a_{ij}))$ for each $n \times n$ matrix $(a_{ij}) \in \mathsf{M}_n(\mathcal{A})$. Subsequently we will use without comment the simple facts that $\mathsf{M}_n(\mathcal{A})$ is a module over $\mathsf{M}_n(\mathbf{C})$ and each ϕ_n is an $\mathsf{M}_n(\mathbf{C})$-module map. We say that ϕ is *completely positive* if each ϕ_n is positive. More generally, "completely" is added to a property of ϕ if it holds for the sequence $\{\phi_n\}_{n=1}^{\infty}$ of maps. Thus ϕ is *completely contractive* if every ϕ_n is contractive, ϕ is *completely isometric* if each ϕ_n is isometric, while ϕ is *completely bounded* if the sequence $\{\|\phi_n\|\}_{n=1}^{\infty}$ is uniformly bounded. In the latter case $\|\phi\|_{cb}$ is defined to be $\sup_n \|\phi_n\|$.

It is easily checked that a $*$-representation $\pi \colon \mathcal{A} \to B(K)$ is completely positive:

$$\pi_n((a_{ij})^*(a_{ij})) = \pi_n(a_{ij})^* \pi_n(a_{ij}) \geq 0$$

for any matrix $(a_{ij}) \in \mathsf{M}_n(\mathcal{A})$, using the algebraic properties of π. For a fixed operator $V \colon H \to K$ the map $\phi \colon \mathcal{A} \to B(H)$ defined by

$$\phi(a) = V^* \pi(a) V, \qquad a \in \mathcal{A},$$

is also completely positive. To see this, let $(a_{ij}) \in \mathsf{M}_n(\mathcal{A})$ be positive and let

$$V_n = \begin{pmatrix} V & & 0 \\ & \ddots & \\ 0 & & V \end{pmatrix}.$$

Then

$$\phi_n(a_{ij}) = V_n^*(\pi(a_{ij}))V_n \geq 0.$$

The most basic result in the subject is Stinespring's representation theorem for completely positive maps, which asserts that we already have the most general form of such maps.

1.2.1 Theorem. *Let $\mathcal{A}$ be a C^*-algebra and let $\phi \colon \mathcal{A} \to B(H)$ be a completely positive map. Then there exists a Hilbert space K, a representation $\pi \colon \mathcal{A} \to B(K)$ and a bounded operator $V \colon H \to K$ such that*

$$\phi(a) = V^* \pi(a) V, \qquad a \in \mathcal{A}.$$

Proof. We first assume that $\mathcal{A}$ is unital with identity 1. On the algebraic tensor product $\mathcal{A} \otimes H$ define a semi-inner product by

$$\left(\sum_{j=1}^{n} a_j \otimes h_j, \sum_{i=1}^{n} b_i \otimes k_i \right) = \sum_{i,j=1}^{n} \langle \phi(b_i^* a_j) h_j, k_i \rangle$$

for $a_j, b_i \in \mathcal{A}$, $h_j, k_i \in H$. Two sums represent the same element of $\mathcal{A} \otimes H$ precisely when they differ by a finite sum of terms

$$((a + b) \otimes h - a \otimes h - b \otimes h) \quad \text{or} \quad (a \otimes (h + k) - a \otimes h - a \otimes k)$$

and so it is easy to verify that the semi-inner product is well defined. If $u = \sum_{i=1}^{n} a_i \otimes h_i \in \mathcal{A} \otimes H$ then

$$(u, u) = \sum_{i,j=1}^{n} \langle \phi(a_i^* a_j) h_j, h_i \rangle$$

$$= \left\langle \phi_n \left[\begin{pmatrix} a_1^* & 0 & \cdots & 0 \\ \vdots & & & \\ a_n^* & 0 & \cdots & 0 \end{pmatrix} \begin{pmatrix} a_1 & \cdots & a_n \\ 0 & \cdots & 0 \\ 0 & \cdots & 0 \end{pmatrix} \right] \begin{pmatrix} h_1 \\ \vdots \\ h_n \end{pmatrix}, \begin{pmatrix} h_1 \\ \vdots \\ h_n \end{pmatrix} \right\rangle \geq 0$$

by the complete positivity of ϕ. It may be the case that $(u, u) = 0$ for some non-zero elements $u \in \mathcal{A} \otimes H$, but nevertheless the Cauchy–Schwarz inequality is still valid:

$$|(u, v)|^2 \leq (u, u)(v, v).$$

Let $N = \{ u \in \mathcal{A} \otimes H : (u, u) = 0 \}$. Then

$$(u, v) = (v, u) = 0$$

for $u \in N, v \in \mathcal{A} \otimes H$. Thus $(\cdot, \cdot)$ induces a well-defined inner product $[\cdot, \cdot]$ on the quotient space $L = (\mathcal{A} \otimes H)/N$ by

$$[u + N, v + N] = (u, v).$$

The C^*-algebra acts on $\mathcal{A} \otimes H$ by left multiplication in the first variable:

$$a \cdot u = \sum_{i=1}^{n} a a_i \otimes h_i$$

where $u = \sum\limits_{i=1}^{n} a_i \otimes h_i$. For any $a_1, \ldots, a_n \in \mathcal{A}$

$$\begin{pmatrix} a_1^* \\ \vdots \\ a_n^* \end{pmatrix} \begin{pmatrix} a^*a & & \\ & \ddots & \\ & & a^*a \end{pmatrix} (a_1, \ldots, a_n) \leq \|a\|^2 (a_j^* a_i)$$

in $\mathbf{M}_n(\mathcal{A})$ and so the complete positivity of ϕ shows that

$$(a \cdot u, a \cdot u) \leq \|a\|^2 (u, u)$$

for $u \in \mathcal{A} \otimes H$. Thus N is invariant under the action of $\mathcal{A}$. We may then define a representation π of $\mathcal{A}$ on L by

$$\pi(a) \left(\sum_{i=1}^{n} a_i \otimes h_i + N \right) = \sum_{i=1}^{n} a a_i \otimes h_i + N.$$

From above

$$[a \cdot u + N, a \cdot u + N] \leq \|a\|^2 [u + N, u + N]$$

and so $\pi(a)$ is a bounded operator satisfying $\|\pi(a)\| \leq \|a\|$. There is then a unique continuous extension, also denoted $\pi(a)$, to the completion K of L. Now define $V \colon H \to K$ by

$$Vh = 1 \otimes h + N, \qquad h \in H.$$

Then
$$\|Vh\|^2 = [1 \otimes h + N, 1 \otimes h + N] = (1 \otimes h, 1 \otimes h)$$
$$= \langle \phi(1)h, h \rangle \leq \|\phi(1)\| \|h\|^2$$

and so $\|V\| \leq \|\phi(1)\|^{1/2}$. For $a \in \mathcal{A}, h_1, h_2 \in H$,

$$\langle V^* \pi(a) V h_1, h_2 \rangle = [a \otimes h_1 + N, 1 \otimes h_2 + N]$$
$$= (a \otimes h_1, 1 \otimes h_2)$$
$$= \langle \phi(a) h_1, h_2 \rangle$$

and so $\phi(a) = V^* \pi(a) V$.

We now turn to the non-unital case, so consider a non-unital C^*-algebra $\mathcal{A}$ and a completely positive map $\phi \colon \mathcal{A} \to B(H)$. If ϕ were unbounded then there would exist a sequence $\{a_n\}_{n=1}^{\infty}$ of positive elements in the unit ball of $\mathcal{A}$ such that $\|\phi(a_n)\| \geq 4^n$. Let $a = \sum\limits_{n=1}^{\infty} 2^{-n} a_n$. Then

$$a \geq 2^{-n} a_n, \qquad n \geq 1,$$

and so $\phi(a) \geq 2^{-n}\phi(a_n)$. Thus

$$\|\phi(a)\| \geq 2^{-n}\|\phi(a_n)\| \geq 2^n, \qquad n \geq 1,$$

which is clearly impossible. Thus ϕ is bounded.

Now fix an increasing positive approximate identity $\{e_\lambda\}_{\lambda\in\Lambda}$, $\|e_\lambda\| \leq 1$, for $\mathcal{A}$ and let $\mathcal{B}$ be the C^*-algebra obtained by adjoining an identity element 1 to $\mathcal{A}$. The net $\{\phi(e_\lambda)\}$ is increasing in $B(H)$, and uniformly bounded by $\|\phi\|$, so we extend ϕ to $\mathcal{B}$ by setting $\phi(1) = \|\phi\|I$. We will show that ϕ is completely positive on $\mathcal{B}$.

Fix $n \geq 1$, and let $E_\lambda \in \mathsf{M}_n(\mathcal{A})$ be the diagonal matrix whose every diagonal entry is e_λ. Then $\{E_\lambda\}_{\lambda\in\Lambda}$ is an approximate identity for $\mathsf{M}_n(\mathcal{A})$. Consider an element

$$A + P \geq 0$$

where $A \in \mathsf{M}_n(\mathcal{A}), P \in \mathsf{M}_n(\mathbf{C})$. Then

$$(I - E_\lambda)A(I - E_\lambda) + P(I - E_\lambda)^2 \geq 0.$$

Since $\{E_\lambda\}_{\lambda\in\Lambda}$ is an approximate identity, given $\varepsilon > 0$ we may choose λ so that the first term of the sum has norm at most ε. Then

$$P(I - E_\lambda)^2 \geq -\varepsilon I.$$

Since P is a scalar matrix, it follows that $P \geq 0$ since $E_\lambda \neq I$. Let $P_k = P + k^{-1}I$, $k \geq 1$, so that P_k is both positive and invertible. Then we may write

$$P_k = U_k^* D_k U_k$$

where U_k is a diagonalizing unitary for P_k and D_k is a diagonal invertible matrix. Since

$$A + P_k \geq A + P \geq 0,$$

it follows that

$$D_k^{-1/2}U_k A U_k^* D_k^{-1/2} + I \geq 0.$$

Pre- and post-multiply by E_λ to obtain

$$D_k^{-1/2}U_k E_\lambda A E_\lambda U_k^* D_k^{-1/2} + E_\lambda^2 \geq 0.$$

This is now an inequality in $\mathsf{M}_n(\mathcal{A})$ so we may apply ϕ_n to obtain

$$D_k^{-1/2}U_k \phi_n(E_\lambda A E_\lambda) U_k^* D_k^{-1/2} + \|\phi\|I \geq 0,$$

since $\|\phi\|I \geq \phi_n(E_\lambda^2)$. Thus, for each $k \geq 1$,

$$\phi_n(E_\lambda A E_\lambda) + \|\phi\|P_k \geq 0.$$

Let $k \to \infty$, when this inequality becomes

$$\phi_n(E_\lambda A E_\lambda) + \|\phi\| P \geq 0.$$

We obtain the positivity of ϕ_n on $\mathbf{M}_n(\mathcal{B})$ by letting $\lambda \to \infty$ since

$$\lim_\lambda \|E_\lambda A E_\lambda - A\| = 0.$$

Thus

$$\phi_n(A) + \|\phi\| P \geq 0.$$

We now have a completely positive map $\phi \colon \mathcal{B} \to B(H)$ so we may apply the first part to the unital C^*-algebra $\mathcal{B}$. There exists a representation $\pi \colon \mathcal{B} \to B(K)$ and a bounded operator $V \colon H \to K$ such that

$$\phi(b) = V^* \pi(b) V, \qquad b \in \mathcal{B}.$$

Now restrict this representation to $\mathcal{A}$.

The proof of this theorem contains some important extra information which we now list.

1.2.2 Remark. (i) If $\phi \colon \mathcal{A} \to B(H)$ is completely positive and $\mathcal{A}$ is unital then ϕ is bounded and

$$\|\phi\| = \|\phi(1)\|.$$

We obtained $\|V\|^2 = \|\phi(1)\| = \|\phi\|$ and so the representation of ϕ is optimal in the sense that no operator of smaller norm could replace V. Moreover it is easy to see that

$$\|\phi_n\| = \|\phi_n(I_n)\| = \|\phi(1)\| = \|\phi\|$$

so that ϕ is completely bounded with $\|\phi\|_{cb} = \|\phi\|$.

(ii) If $\mathcal{A}$ is non-unital then let $\mu = \sup_\lambda \|\phi(e_\lambda)\|$. When extending ϕ from $\mathcal{A}$ to $\mathcal{B}$, the proof would have also been valid if we had defined $\phi(1)$ to be $\mu \leq \|\phi\|$. Then, from the unital case,

$$\|V\|^2 \leq \mu,$$

showing that

$$\|\phi\| \leq \|V\|^2 \leq \mu \leq \|\phi\|.$$

In particular

$$\|\phi\|_{cb} = \|\phi\| = \sup_\lambda \|\phi(e_\lambda)\|$$

in this case.

(iii) If A is a von Neumann algebra and ϕ is ultraweakly continuous then the representation $\pi\colon A \to B(K)$ may be chosen to be ultraweakly continuous. This is immediate from the construction.

The second basic result on this topic is Arveson's extension theorem [Ar] for completely bounded maps. We say that a norm closed subspace $\mathcal{E}$ of a unital C^*-algebra A is an *operator system* if $\mathcal{E}$ is self-adjoint and contains the identity. The matrix spaces $M_n(\mathcal{E})$ are viewed as subspaces of $M_n(A)$, $n \geq 1$, and the ordering in $M_n(\mathcal{E})$ is that which is inherited from $M_n(A)$. The definition of complete positivity extends naturally to this setting.

1.2.3 Theorem. *Let $\mathcal{E}$ be an operator system in a unital C^*-algebra A and let $\phi\colon \mathcal{E} \to B(H)$ be a completely positive unital map. Then ϕ extends to a completely positive map $\psi\colon A \to B(H)$, and $\|\psi\| = \|\phi\| = \|\phi\|_{cb} = \|\psi\|_{cb}$.*

Proof. If $F \in M_n(\mathcal{E})$ then $\begin{pmatrix} 0 & F \\ F^* & 0 \end{pmatrix} \in M_{2n}(\mathcal{E})$ is self-adjoint with norm equal to $\|F\|$. It follows from this that $\|F\| \leq 1$ if and only if $\begin{pmatrix} I & \pm F \\ \pm F^* & I \end{pmatrix} \geq 0$ in $M_{2n}(\mathcal{E})$. Thus, if $\|F\| \leq 1$, then

$$\begin{pmatrix} I & \pm\phi_n(F) \\ \pm\phi_n(F) & I \end{pmatrix} \geq 0$$

and so $\|\phi_n(F)\| \leq 1$. Hence ϕ is completely contractive.

We begin by considering the case in which H is finite dimensional, so that $H = C^k$ for some $k \geq 1$. Let $\{\xi_1, \ldots, \xi_k\}$ be the standard basis for C^k and let $\xi \in C^{k^2}$ be the vector $(\xi_1, \xi_2, \ldots, \xi_k)$. We define a linear functional ω on $M_k(\mathcal{E})$ by

$$\omega((e_{ij})) = \langle \phi_k(e_{ij})\xi, \xi \rangle = \sum_{i,j=1}^{n} \langle \phi(e_{ij})\xi_j, \xi_i \rangle$$

for $(e_{ij}) \in M_k(\mathcal{E})$. Since ϕ_k is contractive,

$$|\omega((e_{ij}))| \leq \|(e_{ij})\|\|\xi\|^2 = k\|(e_{ij})\|$$

while

$$\omega(I_k) = \langle I_k \xi, \xi \rangle = k.$$

Thus ω has norm k which it attains at the identity. By the Hahn–Banach theorem, there is a norm preserving extension to a linear functional θ on $M_k(A)$ which also attains its norm at the identity. Thus θ is a multiple of a state and so is positive.

Let $E_{ij} \otimes a$, for $a \in A$, be the matrix in $M_n(A)$ whose only non-zero entry is a in the (i, j) position. Then define $\psi\colon A \to B(C^k)$ by

$$\langle \psi(a)\xi_j, \xi_i \rangle = \theta(E_{ij} \otimes a), \qquad 1 \leq i, j \leq k.$$

The definition of ω ensures that ψ is an extension of ϕ. We now check that ψ is completely positive. Let (a_{ij}) be a positive matrix in $\mathbf{M}_n(\mathcal{A})$. We must show that

$$\sum_{i,j} \langle \psi(a_{ij})\eta_j, \eta_i \rangle \geq 0$$

for arbitrary vectors $\eta_i \in \mathbf{C}^k$. Write $\eta_i = \sum_p \lambda_{ip}\xi_p$ in terms of the standard basis. Then

$$\sum_{i,j} \langle \psi(a_{ij})\eta_j, \eta_i \rangle = \sum_{i,j,p,q} \lambda_{jq}\bar{\lambda}_{ip} \langle \psi(a_{ij})\xi_q, \xi_p \rangle$$

$$= \sum_{i,j,p,q} \lambda_{jq}\bar{\lambda}_{ip} \theta(E_{pq} \otimes a_{ij}).$$

Any representation of $\mathbf{M}_n \otimes \mathcal{A}$ is unitarily equivalent to one of the form $I \otimes \pi$ on $H \oplus \cdots \oplus H$ where $\pi\colon \mathcal{A} \to B(H)$ is a representation of $\mathcal{A}$. Thus the GNS representation of θ gives $\pi\colon \mathcal{A} \to B(H)$ and vectors $h_1, \ldots, h_n \in H$ such that

$$\theta(E_{ij} \otimes a) = \langle \pi(a)h_j, h_i \rangle.$$

Thus

$$\sum_{i,j} \langle \psi(a_{ij})\eta_j, \eta_i \rangle = \sum_{i,j,p,q} \lambda_{jq}\bar{\lambda}_{ip} \langle \pi(a_{ij})h_q, h_p \rangle$$

$$= \left\langle (\pi(a_{ij})) \begin{pmatrix} h_1' \\ \vdots \\ h_n' \end{pmatrix}, \begin{pmatrix} h_1' \\ \vdots \\ h_n' \end{pmatrix} \right\rangle$$

where $h_i' = \sum_p \lambda_{ip}h_p$. Representations are completely positive so the matrix $(\pi(a_{ij}))$ is positive, and thus $\sum_{ij} \langle \psi(a_{ij})\eta_j, \eta_i \rangle \geq 0$. This shows that ψ is completely positive. We note that ψ is unital, so that $\|\psi\| = \|\phi\|$ by Remark 1.2.2.

Now let H be a general Hilbert space, and let S be the set of finite dimensional subspaces ordered by inclusion. We fix a Banach limit LIM on $\ell^\infty(S)$. This is a state which assigns its limit to any bounded convergent net. If F is finite dimensional then the compression of ϕ to $B(F)$ is completely positive and thus has a unital completely positive extension $\psi_F\colon \mathcal{A} \to B(F)$. Denote the projection of H onto F by P_F and, for each $a \in \mathcal{A}$, define $f_a\colon H \times H \to \mathbf{C}$ by

$$f_a(h_1, h_2) = \mathop{\mathrm{LIM}}_F \langle \psi_F(a)P_F h_1, P_F h_2 \rangle.$$

LIM assigns the same value to any two nets which are eventually identical, and so f_a is a sesquilinear form on $H \times H$. Since

$$|f_a(h_1, h_2)| \leq \|a\|\|h_1\|\|h_2\|,$$

there is an operator, denoted $\psi(a)$, in $B(H)$ such that $\|\psi(a)\| \le \|a\|$ and

$$f_a(h_1, h_2) = \langle \psi(a)h_1, h_2 \rangle.$$

Using the properties of LIM it is easy to verify that $\psi \colon \mathcal{A} \to B(H)$ is a linear map. If $a \in \mathcal{E}$ and F contains $\mathrm{span}\{h_1, h_2\}$ then

$$\langle \psi_F(a)P_F h_1, P_F h_2 \rangle = \langle \phi(a)h_1, h_2 \rangle$$

so that the net $\{\langle \psi_F(a)P_F h_1, P_F h_2 \rangle\}_{F \in \mathcal{S}}$ is eventually constant. Thus, for $a \in \mathcal{E}$,

$$\langle \psi(a)h_1, h_2 \rangle = \mathop{\mathrm{LIM}}_{F} \langle \psi_F(a)P_F h_1, P_F h_2 \rangle$$
$$= \langle \phi(a)h_1, h_2 \rangle,$$

so ψ is an extension of ϕ. Since

$$\sum_{i,j} \langle \psi(a_{ij})h_j, h_i \rangle = \mathop{\mathrm{LIM}}_{F} \sum_{i,j} \langle \psi_F(a_{ij})P_F h_j, P_F h_i \rangle$$

for $a_{ij} \in \mathcal{A}$, $h_i \in H$, the complete positivity of ψ is a consequence of the positivity of LIM and the complete positivity of each ψ_F. The norm equalities are immediate from Remark 1.2.2.

For subsequent developments it will be useful to have an abstract characterization of operator systems. The following definition is motivated by considering the properties of an operator system in a C^*-algebra.

1.2.4 Definition. A complex vector space $\mathcal{V}$ is said to be *matrix ordered* if

(i) $\mathcal{V}$ is a $*$-vector space,
(ii) each matrix space $\mathbf{M}_n(\mathcal{V}), n \ge 1$, is partially ordered by a positive cone $\mathbf{M}_n(\mathcal{V})^+$,
(iii) $S^* \mathbf{M}_m(\mathcal{V})^+ S \subseteq \mathbf{M}_n(\mathcal{V})^+$ for any scalar $m \times n$ matrix S,
(iv) $\mathcal{V}^+ \subseteq \mathcal{V}_h$, the real vector subspace of self-adjoint elements in $\mathcal{V}$.

We note that suitable choices of the matrix S in (iii) show that $\mathbf{M}_n(\mathcal{V})^+$ is contained in $(\mathbf{M}_n(\mathcal{V}))_h$ for each $n \ge 1$. For example, if $\begin{pmatrix} a & b \\ c & d \end{pmatrix} \ge 0$ in $\mathbf{M}_2(\mathcal{V})$ then, taking $S = \begin{pmatrix} \alpha \\ \beta \end{pmatrix}$, we have

$$|\alpha|^2 a + |\beta|^2 d + \bar{\alpha}\beta b + \alpha\bar{\beta}c \ge 0.$$

Setting $\alpha = 0$ and then $\beta = 0$ gives $a, d \in \mathcal{V}_h$, and so

$$\bar{\alpha}\beta b + \alpha\bar{\beta}c \in \mathcal{V}_h.$$

Let $\alpha = 1, \beta = e^{i\theta}$. Then

$$e^{i\theta}b + e^{-i\theta}c \in \mathcal{V}_h$$

so

$$e^{i\theta}b + e^{-i\theta}c = e^{-i\theta}b^* + e^{i\theta}c^*,$$
$$b - c^* = e^{-2i\theta}(b^* - c), \quad \theta \in [0, 2\pi],$$

which can only be true if $b^* = c$. Thus $\begin{pmatrix} a & b \\ c & d \end{pmatrix} \in (\mathsf{M}_2(\mathcal{V}))_h$. It is also clear that $\mathcal{V}_h \cap i\mathcal{V}_h = 0$, and $\mathcal{V} = \mathcal{V}_h \oplus i\mathcal{V}_h$.

A matrix ordered space $\mathcal{V}$ is said to have an *order unit* $e \in \mathcal{V}^+$ if, for each $v \in \mathcal{V}_h$, there exists $t > 0$ such that

$$-te \le v \le te.$$

The ordering is said to be *Archimedean* if the inequality

$$-te \le v, \text{ for all } t > 0$$

implies $v \ge 0$.

1.2.5 Lemma. *Let $\mathcal{V}$ be a matrix ordered space with an order unit e. Then*

$$\begin{pmatrix} e & & \\ & \ddots & \\ & & e \end{pmatrix} \in \mathsf{M}_n(\mathcal{V}) \text{ is an order unit for } \mathsf{M}_n(\mathcal{V}) \text{ for all } n \ge 1.$$

Proof. Consider first the case $n = 2$. Each self-adjoint element of $\mathsf{M}_2(\mathcal{V})$ may be written as a sum

$$\begin{pmatrix} a & 0 \\ 0 & b \end{pmatrix} + \begin{pmatrix} 0 & c \\ c & 0 \end{pmatrix} + \begin{pmatrix} 0 & id \\ -id & 0 \end{pmatrix}$$

with $a, b, c, d \in \mathcal{V}_h$ so it suffices to consider each of the three types of elements separately.

Choose $t > 0$ such that

$$te \pm a, \ te \pm b \ge 0.$$

Then

$$\begin{pmatrix} te \pm a & 0 \\ 0 & te \pm b \end{pmatrix} = \begin{pmatrix} 1 \\ 0 \end{pmatrix}(te \pm a)(1 \ 0) + \begin{pmatrix} 0 \\ 1 \end{pmatrix}(te \pm b)(0 \ 1) \ge 0.$$

Now choose $t > 0$ such that

$$te \pm c \ge 0.$$

Then

$$\begin{pmatrix} te & \pm c \\ \pm c & te \end{pmatrix} = \frac{1}{2}\begin{pmatrix} 1 \\ 1 \end{pmatrix}(te \pm c)(1\ 1) + \frac{1}{2}\begin{pmatrix} 1 \\ -1 \end{pmatrix}(te \mp c)(1\ -1) \geq 0.$$

Finally choose $t > 0$ such that

$$te \pm d \geq 0.$$

Then

$$\begin{pmatrix} te & \pm id \\ \mp id & te \end{pmatrix} = \frac{1}{2}\begin{pmatrix} 1 \\ -i \end{pmatrix}(te \pm d)(1\ i) + \frac{1}{2}\begin{pmatrix} 1 \\ i \end{pmatrix}(te \mp d)(1\ -i) \geq 0.$$

Thus $\begin{pmatrix} e & 0 \\ 0 & e \end{pmatrix}$ is an order unit for $M_2(\mathcal{V})$.

For the general case, observe that an arbitrary self-adjoint matrix in $M_n(\mathcal{V})$ decomposes as a sum of self-adjoint matrices where, for a fixed pair i, j, only the $(i, i), (j, j), (i, j)$ and (j, i) entries are possibly non-zero. For example, in $(M_3(\mathcal{V}))_h$,

$$\begin{pmatrix} a & d & e \\ d^* & b & f \\ e^* & f^* & c \end{pmatrix} = \begin{pmatrix} a & d & 0 \\ d^* & b & 0 \\ 0 & 0 & 0 \end{pmatrix} + \begin{pmatrix} 0 & 0 & e \\ 0 & 0 & 0 \\ e^* & 0 & 0 \end{pmatrix} + \begin{pmatrix} 0 & 0 & 0 \\ 0 & 0 & f \\ 0 & f^* & c \end{pmatrix}.$$

Each such matrix has the form

$$U^* \begin{pmatrix} a & b & & \\ b^* & c & & \\ & & 0 & \\ & & & \ddots \\ & & & & 0 \end{pmatrix} U$$

where $U \in M_n(\mathbf{C})$ is a permutation unitary. From the case $n = 2$, choose $t > 0$ such that

$$\begin{pmatrix} te & 0 \\ 0 & te \end{pmatrix} \pm \begin{pmatrix} a & b \\ b^* & c \end{pmatrix} \geq 0.$$

Then

$$\begin{pmatrix} te & & \\ & \ddots & \\ & & te \end{pmatrix} \pm U^* \begin{pmatrix} a & b & & \\ b^* & c & & \\ & & 0 & \\ & & & \ddots \\ & & & & 0 \end{pmatrix} U$$

$$= U^* \left[\begin{pmatrix} te \pm a & \pm b & & \\ \pm b^* & te \pm c & & \\ & & 0 & \\ & & & \ddots \\ & & & & 0 \end{pmatrix} + \begin{pmatrix} 0 & & & \\ & 0 & & \\ & & te & \\ & & & \ddots \\ & & & & te \end{pmatrix} \right] U \geq 0,$$

since U commutes with $\begin{pmatrix} te \\ & \ddots \\ & & te \end{pmatrix}$. Thus $\begin{pmatrix} e \\ & \ddots \\ & & e \end{pmatrix}$ is an order

unit for $\mathbf{M}_n(\mathcal{V})$.

If $\mathcal{V}$ is an Archimedean matrix ordered space with an order unit e and $\mathcal{V}^+$ is proper, then we may define norms on $\mathbf{M}_n(\mathcal{V})$ by

$$\|A\| = \inf\left\{ \lambda: \ -\begin{pmatrix} \lambda e_n & 0 \\ 0 & \lambda e_n \end{pmatrix} \le \begin{pmatrix} 0 & A \\ A^* & 0 \end{pmatrix} \le \begin{pmatrix} \lambda e_n & 0 \\ 0 & \lambda e_n \end{pmatrix} \right\}$$

where $e_n = \begin{pmatrix} e \\ & \ddots \\ & & e \end{pmatrix} \in \mathbf{M}_n(\mathcal{V})$, and the inequalities take place in

$\mathbf{M}_{2n}(\mathcal{V})$. The triangle inequality is immediate from the definition, while $\|\alpha A\| = |\alpha| \, \|A\|$ is an easy consequence of the matrix equation

$$\begin{pmatrix} \lambda e_n & \pm e^{i\theta} A \\ \pm e^{-i\theta} A^* & \lambda e_n \end{pmatrix} = \begin{pmatrix} e^{i\theta} & 0 \\ 0 & 1 \end{pmatrix} \begin{pmatrix} \lambda e_n & \pm A \\ \pm A^* & \lambda e_n \end{pmatrix} \begin{pmatrix} e^{-i\theta} & 0 \\ 0 & 1 \end{pmatrix}.$$

If $\|A\| = 0$, then for each integer $k \ge 1$,

$$-\begin{pmatrix} e_n & 0 \\ 0 & e_n \end{pmatrix} \le k \begin{pmatrix} 0 & A \\ A^* & 0 \end{pmatrix} \le \begin{pmatrix} e_n & 0 \\ 0 & e_n \end{pmatrix}$$

from the definition. Compression to the (i, i), $(i, j + n)$, $(j + n, i)$ and $(j + n, j + n)$ entries can be achieved by multiplication on the left by a suitable $S \in \mathbf{M}_{2n}$ and on the right by S^*, so

$$-\begin{pmatrix} e & 0 \\ 0 & e \end{pmatrix} \le k \begin{pmatrix} 0 & a_{ij} \\ a_{ij}^* & 0 \end{pmatrix} \le \begin{pmatrix} e & 0 \\ 0 & e \end{pmatrix}$$

in $\mathbf{M}_2(\mathcal{V})$, where a_{ij} is the (i, j) entry of A. It follows that, for arbitrary $\alpha, \beta \in \mathbf{C}$, $k \ge 1$,

$$-(|\alpha|^2 + |\beta|^2)e \le k(\bar{\alpha}\beta a_{ij} + \alpha\bar{\beta} a_{ij}^*) \le (|\alpha|^2 + |\beta|^2)e.$$

Since $\mathcal{V}$ is Archimedean ordered and $\mathcal{V}^+$ is proper, we obtain

$$\bar{\alpha}\beta a_{ij} + \alpha\bar{\beta} a_{ij}^* = 0.$$

Put $\alpha = 1$, $\beta = e^{i\theta}$ to conclude that $a_{ij} = 0$. Thus $A = 0$ and $\|\cdot\|$ does define a norm on $\mathbf{M}_n(\mathcal{V})$.

For self-adjoint elements $v \in \mathcal{V}_h$ the norm may also be expressed by

$$\|v\| = \inf\{\lambda: \ -\lambda e \le v \le \lambda e\}.$$

This is immediate from the relations

$$2(\lambda e \pm v) = (1\ 1) \begin{pmatrix} \lambda e & \pm v \\ \pm v & \lambda e \end{pmatrix} \begin{pmatrix} 1 \\ 1 \end{pmatrix}$$

and

$$\begin{pmatrix} \lambda e & \pm v \\ \pm v & \lambda e \end{pmatrix} = \frac{1}{2} \begin{pmatrix} 1 \\ 1 \end{pmatrix} (\lambda e \pm v)(1\ 1) + \frac{1}{2} \begin{pmatrix} 1 \\ -1 \end{pmatrix} (\lambda e \mp v)(1\ -1)$$

which show that positivity of $\begin{pmatrix} \lambda e & \pm v \\ \pm v & \lambda e \end{pmatrix}$ is equivalent to positivity of $\lambda e \pm v$.

If $\mathcal{V}$ is an Archimedean matrix ordered space with order unit e and proper cone $\mathcal{V}^+$ then we will henceforth assume that the matrix spaces $M_n(\mathcal{V})$ are endowed with the norms defined above. Simple matrix calculations show that the embedding of $\mathcal{V}$ into $M_n(\mathcal{V})$ given by $v \to v \otimes E_{ij}$, for a fixed matrix unit E_{ij}, is isometric and so $M_n(\mathcal{V})^*$ may be identified with $M_n(\mathcal{V}^*)$. If $(\phi_{ij}) \in M_n(\mathcal{V}^*)$ and $(v_{ij}) \in M_n(\mathcal{V})$, the action is

$$(\phi_{ij})(v_{ij}) = \sum_{i,j} \phi_{ij}(v_{ij}).$$

The dual cones are defined by

$$(\phi_{ij}) \in M_n(\mathcal{V})^{*+} \Leftrightarrow \sum_{i,j} \phi_{ij}(v_{ij}) \geq 0 \text{ for all } (v_{ij}) \in M_n(\mathcal{V})^+.$$

The order unit hypothesis implies that $M_n(\mathcal{V})^+$ spans $M_n(\mathcal{V})$ and so each positive cone $M_n(\mathcal{V})^{*+}$ is proper. The *state space* S of $\mathcal{V}^*$ is defined to be $\{\phi \in \mathcal{V}^{*+}: \phi(e) = 1\}$.

1.2.6 Proposition. *Let $\mathcal{V}$ be an Archimedean matrix ordered space with order unit e and a proper cone $\mathcal{V}^+$, and let S be the state space of $\mathcal{V}^*$. Then*

(i) *each $\phi \in S$ is completely positive,*
(ii) *$\mathcal{V}_h$ and $\mathcal{V}^+$ are norm closed,*
(iii) *$v \in \mathcal{V}^+$ if and only if $\phi(v) \geq 0$ for all $\phi \in S$,*
(iv) *$v = 0$ if and only if $\phi(v) = 0$ for all $\phi \in S$,*
(v) *each $M_n(\mathcal{V})^+$ is a proper cone.*

Proof. (i) Suppose $\phi \in S$ and $v = (v_{ij}) \in M_n(\mathcal{V})^+$. Then $(\phi(v_{ij})) \in M_n$ and so is positive if and only if

$$\sum_{i,j} \phi(v_{ij})\bar{\alpha}_j \alpha_i \geq 0$$

for all scalars $\alpha_i \in \mathbb{C}$. Let R be the row matrix $(\alpha_1, \ldots, \alpha_n)$. Then

$$\sum_{i,j} \phi(v_{ij}) \bar{\alpha}_j \alpha_i = \phi(R v R^*) \geq 0.$$

Thus ϕ is completely positive.

(ii) Since

$$\begin{pmatrix} \lambda & v^* \\ v & \lambda \end{pmatrix} = \begin{pmatrix} 0 & 1 \\ 1 & 0 \end{pmatrix} \begin{pmatrix} \lambda & v \\ v^* & \lambda \end{pmatrix} \begin{pmatrix} 0 & 1 \\ 1 & 0 \end{pmatrix}$$

it is clear from the definition of the norm that the map $v \to v^*$ is a conjugate linear isometry, and so $\mathcal{V}_h$ is norm closed.

Now suppose that $v_n \in \mathcal{V}^+$, $n \geq 1$, and $\lim_{n \to \infty} \|v_n - v\| = 0$. From above $v \in \mathcal{V}_h$, so the alternative definition of the norm on $\mathcal{V}_h$ gives a sequence $\{\lambda_n\}_{n=1}^{\infty}$ such that

$$-\lambda_n e \leq v - v_n \leq \lambda_n e$$

and $\lim_{n \to \infty} \lambda_n = 0$. Then

$$-\lambda_n e \leq (v - v_n) + v_n = v.$$

Since the ordering is Archimedean, this implies that $v \in \mathcal{V}^+$ and $\mathcal{V}^+$ is norm closed.

(iii) One direction is clear from the definition of S. If $v_0 \in \mathcal{V}_h$ but $v_0 \notin \mathcal{V}^+$ then the separation form of the Hahn–Banach theorem gives $\psi \in V^*$ such that

$$\mathrm{Re}\ \psi(v_0) < 0, \quad \mathrm{Re}\ \psi(v) \geq 0$$

for $v \in \mathcal{V}^+$, since $\mathcal{V}^+$ is norm closed. Define $\theta \colon \mathcal{V} \to \mathbb{C}$ by

$$\theta(v) = \psi(v) + \overline{\psi(v^*)}, \qquad v \in \mathcal{V}.$$

From (ii), $\|v\| = \|v^*\|$ and so θ is bounded. Moreover, if $v \in \mathcal{V}_h$ then

$$\theta(v) = 2\mathrm{Re}\ \psi(v)$$

and so $\theta \in \mathcal{V}^{*+}$, but $\theta(v_0) < 0$. Then $\theta(e) \neq 0$, otherwise θ would vanish on $\mathcal{V}_h$, so we may define $\phi = \theta(e)^{-1}\theta \in S$ and $\phi(v_0) < 0$.

For the general case consider $v_0 = u_0 + iw_0 \notin \mathcal{V}^+$ where $u_0, w_0 \in \mathcal{V}_h$. We have already settled the case $w_0 = 0$, so assume $w_0 \neq 0$. Then either $w_0 \notin \mathcal{V}^+$ or $-w_0 \notin \mathcal{V}^+$, and without loss of generality assume the first. Then, from above, there exists $\phi \in S$ such that $\phi(w_0) < 0$. Then

$$\mathrm{Im}\ \phi(v_0) = \phi(w_0) \neq 0$$

and so $\phi(v_0) \notin \mathbf{R}^+$.

(iv) Write $v = u + iw$ with $u, w \in \mathcal{V}_h$. If $v \neq 0$ then we may assume that $u \neq 0$, otherwise replace v by iv. We may also assume that $u \notin \mathcal{V}^+$, otherwise replace v by $-v$. Then, from (iii), there exists $\phi \in S$ such that $\phi(u) < 0$ and so $\phi(v) \neq 0$. Thus, if $\phi(v) = 0$ for all $\phi \in S$, it must be the case that $v = 0$. The reverse direction is obvious.

(v) Suppose that $\pm(v_{ij}) \in M_n(\mathcal{V})^+$ and let $\phi \in S$ be arbitrary. By (i) ϕ is completely positive, so $\pm(\phi(v_{ij})) \in \mathbf{M}_n^+$. Thus $(\phi(v_{ij})) = 0$ and so

$$\phi(v_{ij}) = 0, \qquad 1 \leq i, j \leq n, \quad \phi \in S.$$

It follows from (iv) that $(v_{ij}) = 0$, and so $\mathbf{M}_n(\mathcal{V})^+$ is a proper cone.

We can now establish the characterization of operator systems.

1.2.7 Theorem. *Let $\mathcal{V}$ be a matrix ordered space. Then $\mathcal{V}$ is completely order isomorphic to an operator system if and only if*

(i) $\mathcal{V}^+$ is a proper cone with an order unit e,
(ii) the ordering on each $\mathbf{M}_n(\mathcal{V})$, $n \geq 1$, is Archimedean.

If these conditions are satisfied, the complete order isomorphism θ may be chosen to satisfy $\theta(e) = I$.

Proof. The necessity of these conditions is clear, so suppose that $\mathcal{V}$ is a matrix ordered space satisfying (i) and (ii). Define Ω to be the set of completely positive maps

$$\omega \colon \mathcal{V} \to B(H_\omega)$$

where H_ω is a finite dimensional Hilbert space and $\omega(e) = I$. The set Ω is non-empty because all states on $\mathcal{V}$ are completely positive, by Proposition 1.2.6 (i). If $v \in \mathcal{V}_h$ then, for some $\lambda > 0$,

$$-\lambda e \leq v \leq \lambda e$$

and so

$$-\lambda I \leq \omega(v) \leq \lambda I.$$

Then $\sup_{\omega \in \Omega} \|\omega(v)\|$ is finite. Thus we may form $H = \bigoplus_{\omega \in \Omega} H_\omega$ and define a completely positive map $\theta \colon \mathcal{V} \to B(H)$ by

$$\theta(v) = \bigoplus_{\omega \in \Omega} \omega(v), \qquad v \in \mathcal{V}.$$

Clearly $\theta(e) = I$, and θ is injective by Proposition 1.2.6 (i), (iv). This map will be a complete order isomorphism between $\mathcal{V}$ and its range if we can show that $(v_{ij}) \in M_n(\mathcal{V})^+$ if $(\theta(v_{ij})) \geq 0$. Thus consider an element $(v_{ij}) \notin M_n(\mathcal{V})^+$.

Since the ordering on $M_n(\mathcal{V})$ is Archimedean, the argument of Proposition 1.2.6 (ii) applies to $M_n(\mathcal{V})$, showing that $M_n(\mathcal{V})^+$ is norm closed. The last part of the same proposition shows that $M_n(\mathcal{V})^+$ is proper, and so the separation form of the Hahn–Banach theorem gives $(\phi_{ij}) \in M_n(\mathcal{V}^*)^+$ so that

$$\sum_{i,j} \phi_{ij}(v_{ij}) \notin \mathbf{R}^+.$$

Now define $\alpha\colon \mathcal{V} \to M_n = B(\mathbf{C}^n)$ by

$$\alpha(v) = (\phi_{ij}(v)), \qquad v \in \mathcal{V}.$$

We will show that α is completely positive. To this end, let $w = (w_{pq}) \in M_k(\mathcal{V})^+$. Then $\alpha_k(w) \geq 0$ in $M_k(M_n)$ if for all choices of vectors $\eta_1, \ldots, \eta_k \in \mathbf{C}^n$ we have

$$\sum_{p,q} \langle \alpha(w_{pq})\eta_q, \eta_p \rangle \geq 0.$$

We may write $\eta_p = \sum_i \beta_{pi}\xi_i$, where $\{\xi_i\}_{i=1}^n$ is the standard basis for $\mathbf{C}^n$, and form the matrix $S \in M_{k,n}$ whose (p,i) entry is β_{pi}. Then $S^*wS \in M_n(\mathcal{V})^+$ and its (i,j) entry is $\sum_{pq} \bar{\beta}_{pi}w_{pq}\beta_{qj}$. Since $(\phi_{ij}) \in M_n(\mathcal{V}^*)^+$, it follows that

$$\sum_{ijpq} \bar{\beta}_{pi}\phi_{ij}(w_{pq})\beta_{qj} \geq 0.$$

However

$$\phi_{ij}(w_{pq}) = \langle \alpha(w_{pq})\xi_j, \xi_i \rangle.$$

Thus

$$\sum_{pq} \langle \alpha(w_{pq})\eta_q, \eta_p \rangle = \sum_{ijpq} \langle \alpha(w_{pq})\beta_{qj}\xi_j, \beta_{pi}\xi_i \rangle$$

$$= \sum_{ijpq} \bar{\beta}_{pi}\phi_{ij}(w_{pq})\beta_{qj} \geq 0,$$

showing that α is completely positive. The matrix $(\alpha(v_{ij}))$ is not positive since, if it were, we would have

$$\sum_{ij} \phi_{ij}(v_{ij}) = \sum_{ij} \langle \alpha(v_{ij})\xi_j, \xi_i \rangle \geq 0,$$

a contradiction.

Let $b = \alpha(e) \in M_n^+$. Choose a state μ on $\mathcal{V}$, and replace α by the completely positive map $\alpha_\varepsilon(v) = \alpha(v) + \varepsilon\mu(v)I$. For ε sufficiently small

$\alpha_\varepsilon(e)$ is invertible and $(\alpha_\varepsilon(v_{ij})) \notin \mathsf{M}_{n^2}^+$, and so we may assume that b is invertible. Then define $\omega\colon \mathcal{V} \to \mathsf{M}_n$ by

$$\omega(v) = b^{-1/2}\alpha(v)b^{-1/2}, \qquad v \in \mathcal{V}.$$

Then ω is completely positive, $\omega(e) = I$, and $(\omega(v_{ij})) \notin \mathsf{M}_{n^2}^+$. Thus $(\theta(v_{ij})) \notin \mathsf{M}_n(B(H))^+$ and we have shown that θ is a complete order isomorphism.

1.3 Completely Bounded Maps

The theorems of the previous section have analogues for completely bounded maps, which we now discuss. The two theories are close to one another; some matrix calculations will allow us to reduce to the completely positive situation. We will work in the following setting: $\mathcal{E}$ is an operator space (a subspace of a C^*-algebra $\mathcal{A}$ which we do not assume to be norm closed) and $\phi\colon \mathcal{E} \to B(H)$ is a completely bounded map. It is important to note that $\phi_n\colon \mathsf{M}_n(\mathcal{E}) \to \mathsf{M}_n(B(H))$ is an M_n-bimodule map.

1.3.1 Theorem. *Let $\phi\colon \mathcal{E} \to B(H)$ be completely bounded. Then there exists a representation $\pi\colon \mathcal{A} \to B(K)$ and operators $V, W\colon H \to K$ such that $\|V\| = \|W\| = \|\phi\|_{cb}^{1/2}$ and*

$$\phi(a) = V^*\pi(a)W, \qquad a \in \mathcal{E}.$$

Proof. We assume that $\|\phi\|_{cb} = 1$ since the general case is then obtained by a scaling argument. We introduce in $\mathsf{M}_2(\mathcal{A})$ an operator system $\mathcal{L}$ defined by

$$\mathcal{L} = \left\{ \begin{pmatrix} \lambda & e \\ f^* & \mu \end{pmatrix} : e, f \in \mathcal{E}, \lambda, \mu \in \mathbb{C} \right\}$$

and a map $\theta\colon \mathcal{L} \to \mathsf{M}_2(B(H))$ given by

$$\theta\begin{pmatrix} \lambda & e \\ f^* & \mu \end{pmatrix} = \begin{pmatrix} \lambda & \phi(e) \\ \phi(f)^* & \mu \end{pmatrix}.$$

A typical self-adjoint element of $\mathsf{M}_n(\mathcal{L}) \subseteq \mathsf{M}_{2n}(\mathcal{A})$ may be written

$$\begin{pmatrix} S & X \\ X^* & T \end{pmatrix}$$

with $S, T \in \mathsf{M}_n(\mathbb{C})$, $X \in \mathsf{M}_n(\mathcal{E})$. To show that θ_n is positive, we must verify that $\theta_n \begin{pmatrix} S & X \\ X^* & T \end{pmatrix} \geq 0$ whenever $\begin{pmatrix} S & X \\ X^* & T \end{pmatrix} \geq 0$. Clearly S and T will be positive so it suffices to consider the case when S and T are invertible, since otherwise we may add εI to both and later let ε tend to 0.

Now if $S = T = I$ and $\begin{pmatrix} I & X \\ X^* & I \end{pmatrix} \geq 0$ then, conjugating by the unitary $\begin{pmatrix} I & 0 \\ 0 & -I \end{pmatrix}$, we see that $\begin{pmatrix} I & \pm X \\ \pm X^* & I \end{pmatrix} \geq 0$ and this is equivalent to $\|X\| \leq 1$. Since ϕ is completely contractive, $\|\phi_n(X)\| \leq 1$ and so

$$\theta_n \begin{pmatrix} I & X \\ X^* & I \end{pmatrix} = \begin{pmatrix} I & \phi_n(X) \\ \phi_n(X)^* & I \end{pmatrix} \geq 0.$$

The matrix factorization

$$\begin{pmatrix} S^{1/2} & 0 \\ 0 & T^{1/2} \end{pmatrix} \begin{pmatrix} I & S^{-1/2}XT^{-1/2} \\ T^{-1/2}X^*S^{-1/2} & I \end{pmatrix} \begin{pmatrix} S^{1/2} & 0 \\ 0 & T^{1/2} \end{pmatrix}$$

of $\begin{pmatrix} S & X \\ X^* & T \end{pmatrix}$ shows that $\begin{pmatrix} S & X \\ X^* & T \end{pmatrix} \geq 0$ if and only if

$$\|S^{-1/2}XT^{-1/2}\| \leq 1.$$

Then

$$\phi_n(S^{-1/2}XT^{-1/2}) = S^{-1/2}\phi_n(X)T^{-1/2}$$

and so $\|S^{-1/2}\phi_n(X)T^{-1/2}\| \leq 1$. Then the matrix $\theta_n \begin{pmatrix} S & X \\ X^* & T \end{pmatrix}$ factors as

$$\begin{pmatrix} S^{1/2} & 0 \\ 0 & T^{1/2} \end{pmatrix} \begin{pmatrix} I & S^{-1/2}\phi_n(X)T^{-1/2} \\ T^{-1/2}\phi_n(X)^*S^{-1/2} & I \end{pmatrix} \begin{pmatrix} S^{1/2} & 0 \\ 0 & T^{1/2} \end{pmatrix},$$

and so is non-negative. Thus each θ_n is positive and so θ is completely positive.

By Theorem 1.2.3, θ extends to a completely positive unital map

$$\psi: M_2(\mathcal{A}) \to M_2(B(H)).$$

By Theorem 1.2.1, there exists a representation $\pi: \mathcal{A} \to B(K)$ and a contraction $U: H \oplus H \to K \oplus K$ such that, for $(a_{ij}) \in M_2(\mathcal{A})$,

$$\psi \begin{pmatrix} a_{11} & a_{12} \\ a_{21} & a_{22} \end{pmatrix} = U^* \begin{pmatrix} \pi(a_{11}) & \pi(a_{12}) \\ \pi(a_{21}) & \pi(a_{22}) \end{pmatrix} U.$$

In particular

$$\begin{pmatrix} 0 & \phi(a) \\ 0 & 0 \end{pmatrix} = U^* \begin{pmatrix} 0 & \pi(a) \\ 0 & 0 \end{pmatrix} U, \qquad a \in \mathcal{E},$$

and so

$$\phi(a) = (I, 0)U^* \begin{pmatrix} I \\ 0 \end{pmatrix} \pi(a)(0, I)U \begin{pmatrix} 0 \\ I \end{pmatrix}, \qquad a \in \mathcal{E}.$$

Thus

$$\phi(a) = V^* \pi(a)W, \qquad a \in \mathcal{E},$$

where $V = (I, 0)U \begin{pmatrix} I \\ 0 \end{pmatrix}$ and $W = (0, I)U \begin{pmatrix} 0 \\ I \end{pmatrix}$ are contractions.

We note, from the construction, that the dimension of K may be chosen to be no greater than $\text{card}(\mathcal{A}) \cdot \dim H$ in the infinite dimensional case.

1.3.2 Corollary. *Let $\mathcal{E} \subseteq \mathcal{F}$ be operator spaces in a C^*-algebra $\mathcal{A}$. Then a completely bounded map $\phi \colon \mathcal{E} \to B(H)$ has a norm preserving extension to a completely bounded map $\psi \colon \mathcal{F} \to B(H)$.*

Proof. By scaling, it suffices to consider a complete contraction. Then Theorem 1.3.1 gives a representation

$$\phi(a) = V^* \pi(a)W, \qquad a \in \mathcal{E},$$

where $\|V\|, \|W\| \leq 1$ and π is a representation of $\mathcal{A}$. Define $\psi \colon \mathcal{F} \to B(H)$ by

$$\psi(a) = V^* \pi(a)W, \qquad a \in \mathcal{F}.$$

Clearly ψ extends ϕ, and the estimate $\|\psi\|_{cb} \leq \|V^*\| \|W\| \leq 1$ is easy to obtain.

1.4 The Haagerup Tensor Product

The study of cohomology involves calculations with multilinear maps and so it is important to have useful representations of such maps. Theorem 1.3.1 is a start in this direction, and we will consider bilinear and multilinear maps in the next section. Multilinear maps can be linearized by considering tensor products, the particular tensor product norm being chosen to be compatible with continuity properties of the maps. In the case of completely bounded multilinear maps the appropriate one is the Haagerup tensor product.

If $\mathcal{A}$ is a C^*-algebra then $\mathsf{M}_{m,n}(\mathcal{A})$ will denote the space of rectangular $m \times n$ matrices with entries from $\mathcal{A}$, normed as a subspace of $\mathsf{M}_r(\mathcal{A})$ where $r = \max\{m, n\}$. More generally, if $\mathcal{E}$ is an operator subspace of $\mathcal{A}$, then $\mathsf{M}_{m,n}(\mathcal{E})$ is normed as a subspace of $\mathsf{M}_{m,n}(\mathcal{A})$. Let $\mathcal{E}$ and $\mathcal{F}$ be subspaces of C^*-algebras $\mathcal{A}$ and $\mathcal{B}$. If $E = (e_{ij}) \in \mathsf{M}_{m,r}(\mathcal{E})$ and $F = (f_{ij}) \in \mathsf{M}_{r,n}(\mathcal{F})$ then $E \odot F$ will denote the element of $\mathsf{M}_{m,n}(\mathcal{E} \otimes \mathcal{F})$ whose (i, j) entry is $\sum_k e_{ik} \otimes f_{kj}$. The Haagerup norm on $\mathcal{E} \otimes \mathcal{F}$ is defined by

$$\|u\|_h = \inf\{\|E\| \|F\| \colon u = E \odot F, E \in \mathsf{M}_{1,n}(\mathcal{E}), F \in \mathsf{M}_{n,1}(\mathcal{F})\}.$$

This is equivalent to

$$\|u\|_h = \inf \left\{ \left\| \sum_{i=1}^{n} e_i e_i^* \right\|^{1/2} \left\| \sum_{i=1}^{n} f_i^* f_i \right\|^{1/2} : u = \sum_{i=1}^{n} e_i \otimes f_i \right\}.$$

It is not obvious that this is a norm, but we will establish this fact below.

A fixed element u may have many different representations as finite sums of elementary tensors. We first show that the infimum in the definition is attained.

1.4.1 Lemma. *If $u \in \mathcal{E} \otimes \mathcal{F}$ and $u \neq 0$ then there exist $E \in \mathsf{M}_{1,k}(\mathcal{E})$ and $F \in \mathsf{M}_{k,1}(\mathcal{F})$ such that $u = E \odot F$ and*

$$\|u\|_h = \|E\| \, \|F\|.$$

Moreover, E and F may be chosen to have linearly independent sets of components.

Proof. Let $u = A \odot B, A \in \mathsf{M}_{1,n}(\mathcal{E}), B \in \mathsf{M}_{n,1}(\mathcal{F})$ be any representation of u. Let $\{a_1, \ldots, a_n\}, \{b_1, \ldots, b_n\}$ be the components of A and B respectively. If there are non-trivial linear dependencies among either set suppose, without loss of generality, that one occurs for $\{b_i\}_{i=1}^{n}$. Let $\sum_{i=0}^{n} \lambda_i b_i = 0$ be such a linear dependency, normalized so that $\sum |\lambda_i|^2 = 1$. Then form a unitary matrix $V \in \mathsf{M}_n(\mathbb{C})$ whose last row is $(\lambda_1, \ldots, \lambda_n)$. It is easy to check that u is also represented by $AV^* \odot VB$. Let A_1 and B_1 be respectively AV^* and VB with their last components removed. Since the last component of VB is 0 by construction, $u = A_1 \odot B_1, A_1 \in \mathsf{M}_{1,n-1}(\mathcal{E}), B_1 \in \mathsf{M}_{n-1,1}(\mathcal{F})$ and $\|A_1\| = \|A\|, \|B_1\| = \|B\|$.

Repeat this argument until the following situation is reached: $u = E_1 \odot F_1$, $\|E_1\| = \|A\|, \|F_1\| = \|B\|$, the components $\{e_i\}_{i=1}^{k}$ of E_1 and $\{f_i\}_{i=1}^{k}$ are linearly independent sets. Starting from any other representation $u = C \odot D$, we may also reduce to $u = E_2 \odot F_2$ with $\|E_2\| = \|C\|, \|F_2\| = \|D\|$, where the entries $\{\tilde{e}_i\}_{i=1}^{\ell}$ of E_2 and $\{\tilde{f}_i\}_{i=1}^{\ell}$ of F_2 are linearly independent sets. Choose linear functionals $\phi_i \in \mathcal{E}^*, 1 \leq i \leq k$, and $\psi_i \in \mathcal{E}^*, 1 \leq i \leq \ell$, such that

$$\phi_i(e_j) = \delta_{ij}, \qquad \psi_i(\tilde{e}_j) = \delta_{ij}.$$

Since $u = \sum_{i=1}^{k} e_i \otimes f_i = \sum_{i=1}^{\ell} \tilde{e}_i \otimes \tilde{f}_i$, we obtain

$$f_i = \sum_{j=1}^{\ell} \phi_i(\tilde{e}_j)\tilde{f}_j, \qquad \tilde{f}_i = \sum_{j=1}^{k} \psi_i(e_j)f_j.$$

Thus $\mathrm{span}\{f_1,\ldots,f_k\} = \mathrm{span}\{\tilde{f}_1,\ldots,\tilde{f}_\ell\}$, so in particular $k = \ell$, and both sets of elements are bases. In the same way $\mathrm{span}\{e_1,\ldots,e_k\} = \mathrm{span}\{\tilde{e}_1,\ldots,\tilde{e}_k\}$. Then we may find an invertible matrix $S \in \mathsf{M}_k$ such that $F_2 = SF_1$ and $E_2 = E_1 S^{-1}$.

The representation $u = C \odot D$ was arbitrary, and so

$$\|u\|_h = \inf\{\|E_1 S^{-1}\|\,\|SF_1\|\colon\ S \in \mathsf{M}_k, S \text{ invertible}\}.$$

Replacing S by λS and S^{-1} by S^{-1}/λ for a suitable scalar λ, we may assume that $\|S\| = \|S^{-1}\|$. For such matrices, it is clear that

$$\|E_1 S^{-1}\|\,\|SF_1\| \to \infty \quad \text{as} \quad \|S\| \to \infty$$

so we may restrict to some bounded set given by

$$\|S\|, \|S^{-1}\| \le K, \qquad K > 0.$$

A compactness argument then shows that the infimum defining $\|u\|_h$ is attained.

1.4.2 Proposition. *The map $\|\cdot\|_h$ is a norm on $\mathcal{E} \odot \mathcal{F}$.*

Proof. If $u \ne 0$, then by Lemma 1.4.1 there exists a representation $u = E \odot F$ with $\|u\|_h = \|E\|\,\|F\|$. Clearly $E, F \ne 0$, so $\|u\|_h \ne 0$.

The equation $\|\lambda u\|_h = |\lambda|\,\|u\|_h$ is obvious for $\lambda = 0$, so assume $\lambda \ne 0$. Let $u = E \odot F$ be a representation of u satisfying $\|u\|_h = \|E\|\,\|F\|$. Then $\lambda u = (\lambda E) \odot F$ is a representation of λu. Thus

$$\|\lambda u\|_h \le \|\lambda E\|\,\|F\| = |\lambda|\,\|u\|_h.$$

Then

$$|\lambda|\,\|u\|_h = |\lambda|\,\|\lambda^{-1}(\lambda u)\|_h \le |\lambda|\,|\lambda|^{-1}\|\lambda u\|_h = \|\lambda u\|_h,$$

establishing the reverse inequality.

Now consider a sum $u + v \in \mathcal{E} \otimes \mathcal{F}$, and let $u = E_1 \odot F_1$, $v = E_2 \odot F_2$ be representations such that

$$\|u\|_h = \|E_1\|\,\|F_1\|, \quad \|v\|_h = \|E_2\|\,\|F_2\|,$$

$$E_1 \in \mathsf{M}_{1,n}(\mathcal{E}), E_2 \in \mathsf{M}_{1,r}(\mathcal{E}), \quad F_1 \in \mathsf{M}_{n,1}(\mathcal{F}), F_2 \in \mathsf{M}_{r,1}(\mathcal{F}).$$

We may assume $u, v \ne 0$ otherwise $\|u + v\|_h \le \|u\|_h + \|v\|_h$ is trivial. Let

$$E_\lambda = (\lambda E_1, \lambda^{-1}E_2) \in \mathsf{M}_{1,n+r}(\mathcal{E}), \quad F_\lambda = \begin{pmatrix} \lambda^{-1}F_1 \\ \lambda F_2 \end{pmatrix} \in \mathsf{M}_{n+r,1}(\mathcal{F}), \quad \lambda > 0.$$

Then $u + v$ is represented by $E_\lambda \odot F_\lambda$. Thus

$$
\begin{aligned}
\|u + v\|_h^2 &\leq \|E_\lambda\|^2 \|F_\lambda\|^2 \\
&\leq (\lambda^2 \|E_1\|^2 + \lambda^{-2} \|E_2\|^2)(\lambda^{-2} \|F_1\|^2 + \lambda^2 \|F_2\|^2) \\
&= \|u\|_h^2 + \|v\|_h^2 + \lambda^4 \|E_1\|^2 \|F_2\|^2 + \lambda^{-4} \|E_2\|^2 \|F_1\|^2.
\end{aligned}
$$

Now choose λ such that $\lambda^4 = \|E_2\| \|F_1\| (\|E_1\| \|F_2\|)^{-1}$. This gives

$$
\|u + v\|_h^2 \leq \|u\|_h^2 + \|v\|_h^2 + 2\|u\|_h \|v\|_h
$$

and so $\|u + v\|_h \leq \|u\|_h + \|v\|_h$. This shows that $\| \cdot \|_h$ is a norm.

The completion of $\mathcal{E} \otimes \mathcal{F}$ in this norm is denoted $\mathcal{E} \otimes_h \mathcal{F}$, while the uncompleted algebraic tensor product endowed with the Haagerup norm is written $\mathcal{E} \otimes^h \mathcal{F}$. We follow this convention for any tensor product norm $\| \cdot \|_\alpha$. If $\mathcal{E}_1 \subseteq \mathcal{E}_2$ and $\mathcal{F}_1 \subseteq \mathcal{F}_2$ there is always a natural embedding of $\mathcal{E}_1 \otimes^\alpha \mathcal{F}_1$ into $\mathcal{E}_2 \otimes^\alpha \mathcal{F}_2$ (which may be unbounded). If this embedding is always isometric then we say that $\| \cdot \|_\alpha$ is *injective*.

1.4.3 Proposition. *The Haagerup norm is injective.*

Proof. Suppose $\mathcal{E}_1 \subseteq \mathcal{E}_2$ and $\mathcal{F}_1 \subseteq \mathcal{F}_2$ are operator spaces and $u \in \mathcal{E}_1 \otimes \mathcal{F}_1$. Let λ_1 and λ_2 denote the Haagerup norms of u in $\mathcal{E}_1 \otimes^h \mathcal{F}_1$ and $\mathcal{E}_2 \otimes^h \mathcal{F}_2$ respectively. Any representation of u in $\mathcal{E}_1 \otimes \mathcal{F}_1$ is also a representation of u in $\mathcal{E}_2 \otimes \mathcal{F}_2$ so, from the definition, it is clear that $\lambda_2 \leq \lambda_1$. From Lemma 1.4.1 there exists a representation $u = E_2 \odot F_2$, $E_2 \in M_{1,n}(\mathcal{E}_2)$, $F_2 \in M_{n,1}(\mathcal{F}_2)$, such that $\lambda_2 = \|E_2\| \|F_2\|$ and the entries $\{e_i\}$ of E_2 and $\{f_i\}$ of F_2 form linearly independent sets of elements. There is also a representation $u = \sum_{i=1}^{k} \tilde{e}_i \otimes \tilde{f}_i$ with $\{\tilde{e}_i\}$ independent in $\mathcal{E}_1$ and $\{\tilde{f}_i\}$ independent in $\mathcal{F}_1$. As in the proof of Lemma 1.4.1, this forces

$$
\operatorname{span}\{e_i\} = \operatorname{span}\{\tilde{e}_i\}, \quad \operatorname{span}\{f_i\} = \operatorname{span}\{\tilde{f}_i\}
$$

and so $u = E_2 \odot F_2$ is a representation of u in $\mathcal{E}_1 \otimes^h \mathcal{F}_1$. Thus

$$
\lambda_1 \leq \|E_2\| \|F_2\| = \lambda_2
$$

proving the reverse inequality.

We now introduce norms on $M_n(\mathcal{E} \otimes \mathcal{F})$ for any pair of operator spaces. Any element $U \in M_n(\mathcal{E} \otimes \mathcal{F})$ may be written $U = E \odot F$ for appropriately chosen elements $E \in M_{n,k}(\mathcal{E})$, $F \in M_{k,n}(\mathcal{F})$ for some k and so we may define, for $U \in M_n(\mathcal{E} \otimes \mathcal{F})$,

$$
\|U\|_h = \inf\{\|E\| \|F\| \colon E \in M_{n,k}(\mathcal{E}), F \in M_{k,n}(\mathcal{F}), U = E \odot F\}.
$$

If u_{ij} is the (i, j) entry of U, $E_i \in M_{1,k}(\mathcal{E})$, $F_j \in M_{k,1}(\mathcal{F})$ are respectively the ith row of E and the jth column of F, then

$$u_{ij} = E_i \odot F_j,$$

so $\|u_{ij}\|_h \leq \|E\| \|F\|$, leading to

$$\|u_{ij}\|_h \leq \|U\|_h.$$

Consequently $\|U\|_h = 0$ implies that each entry is 0 and so $U = 0$. The verification that $\| \cdot \|_h$ is a norm on $M_n(\mathcal{E} \otimes \mathcal{F})$ follows the argument of Proposition 1.4.2 and we will not repeat it. As before $M_n(\mathcal{E} \otimes^h \mathcal{F})$ will denote the uncompleted space $M_n(\mathcal{E} \otimes \mathcal{F})$ with the norm just defined, while $M_n(\mathcal{E} \otimes_h \mathcal{F})$ will denote the completion.

Although we now have norms on the matrix spaces over $\mathcal{E} \otimes_h \mathcal{F}$, it is far from obvious that $\mathcal{E} \otimes_h \mathcal{F}$ may be regarded as an operator space. We will construct an embedding of $\mathcal{E} \otimes_h \mathcal{F}$ into an operator algebra in such a way that certain important results can be readily deduced. If the object were to just obtain the embedding, then simpler ways of achieving this are available.

Fix operator subspaces $\mathcal{E}$ and $\mathcal{F}$ of unital C^*-algebras $\mathcal{A}$ and $\mathcal{B}$ respectively. The sets of adjoints of elements in $\mathcal{E}$ and $\mathcal{F}$ form new operator spaces denoted by $\mathcal{E}^*$ and $\mathcal{F}^*$ respectively. Now form the vector space $\mathcal{L}$ whose elements are 2×2 matrices

$$\begin{pmatrix} a & u \\ v & b \end{pmatrix}, \quad a \in \mathcal{A}, \quad b \in \mathcal{B}, \quad u \in \mathcal{E} \otimes \mathcal{F}, \quad v \in \mathcal{F}^* \otimes \mathcal{E}^*.$$

We introduce an adjoint operation from $\mathcal{E} \otimes \mathcal{F}$ to $\mathcal{F}^* \otimes \mathcal{E}^*$ by

$$\left(\sum e_i \otimes f_i \right)^* = \sum f_i^* \otimes e_i^*, \quad e_i \in \mathcal{E}, f_i \in \mathcal{F},$$

which allows us to define an involution on $\mathcal{L}$ by

$$\begin{pmatrix} a & u \\ v & b \end{pmatrix}^* = \begin{pmatrix} a^* & v^* \\ u^* & b^* \end{pmatrix}.$$

The matrix space $M_n(\mathcal{L})$ is then identified with the vector space whose elements have the form

$$\begin{pmatrix} A & U \\ V & B \end{pmatrix}, A \in M_n(\mathcal{A}), B \in M_n(\mathcal{B}), U \in M_n(\mathcal{E} \otimes \mathcal{F}), V \in M_n(\mathcal{F}^* \otimes \mathcal{E}^*).$$

We define an element $\begin{pmatrix} A & U \\ U^* & B \end{pmatrix}$ to be positive if, given $\varepsilon > 0$, there exists a representation $U = E \odot F$, $E \in M_{n,k}(\mathcal{E})$, $F \in M_{k,n}(\mathcal{F})$, such that

$$EE^* \leq A + \varepsilon, \qquad F^*F \leq B + \varepsilon.$$

Note that these inequalities take place in $\mathbf{M}_n(\mathcal{A})$ and $\mathbf{M}_n(\mathcal{B})$ respectively, and imply that $A \geq 0$, $B \geq 0$. Consequently an element which is simultaneously positive and negative must have the form $\begin{pmatrix} 0 & U \\ U^* & 0 \end{pmatrix}$. But then, for any $\varepsilon > 0$, there exists a representation $U = E \odot F$ such that

$$EE^* \leq \varepsilon, \qquad F^*F \leq \varepsilon.$$

Then $\|E\|\,\|F\| \leq \varepsilon$, showing that $\|U\|_h \leq \varepsilon$ for all $\varepsilon > 0$. Thus $U = 0$ and each positive cone is proper. It is easy to check that the cones $\mathbf{M}_n(\mathcal{L})^+$ are also convex.

1.4.4 Proposition. *The space $\mathcal{L}$ is a matrix ordered space, the element $\begin{pmatrix} 1 & 0 \\ 0 & 1 \end{pmatrix}$ is an order unit for $\mathcal{L}$, and each of the cones $\mathbf{M}_n(\mathcal{L})^+$ is Archimedean.*

Proof. Consider a self-adjoint element $\begin{pmatrix} a & u \\ u^* & b \end{pmatrix} \in \mathcal{L}$ where $a \in \mathcal{A}$, $b \in \mathcal{B}$, $u \in \mathcal{E} \otimes \mathcal{F}$, and let $u = \sum_{i=1}^{k} e_i \odot f_i$ be a representation of u. Then let

$$t = \|a\| + \|b\| + \left\| \sum_{i=1}^{k} e_i e_i^* \right\| + \left\| \sum_{i=1}^{k} f_i^* f_i \right\|.$$

For each $\varepsilon > 0$,

$$\sum e_i e_i^* \pm a \leq t \leq t + \varepsilon,$$
$$\sum f_i^* f_i \pm b \leq t \leq t + \varepsilon,$$

and thus

$$\begin{pmatrix} t & 0 \\ 0 & t \end{pmatrix} \geq \pm \begin{pmatrix} a & u \\ u^* & b \end{pmatrix}$$

by definition of the ordering. This shows that $\begin{pmatrix} 1 & 0 \\ 0 & 1 \end{pmatrix}$ is an order unit.

Now suppose that

$$\begin{pmatrix} A & U \\ U^* & B \end{pmatrix} + r \begin{pmatrix} 1 & 0 \\ 0 & 1 \end{pmatrix} \geq 0$$

in $\mathbf{M}_n(\mathcal{L})$ for all $r > 0$. Given $\varepsilon > 0$ choose $r = \varepsilon/2$, so that

$$\begin{pmatrix} A + \varepsilon/2 & U \\ U^* & B + \varepsilon/2 \end{pmatrix} \geq 0.$$

Then there exists a representation $U = E \odot F$ such that

$$EE^* \le (A + \varepsilon/2) + \varepsilon/2 = A + \varepsilon, \qquad F^*F \le (B + \varepsilon/2) + \varepsilon/2 = B + \varepsilon.$$

Thus $\begin{pmatrix} A & U \\ U^* & B \end{pmatrix} \ge 0$ and the ordering is Archimedean.

To show that $\mathcal{L}$ is a matrix ordered space (see the previous section) we need only prove that

$$X^* \mathsf{M}_m(\mathcal{L})^+ X \subseteq \mathsf{M}_n(\mathcal{L})^+$$

for every scalar matrix $X \in \mathsf{M}_{m,n}(\mathbf{C})$. Let $\begin{pmatrix} A & U \\ U^* & B \end{pmatrix} \in \mathsf{M}_m(\mathcal{L})^+$. Given $\varepsilon > 0$, there exists a representation $U = E \odot F$, $E \in \mathsf{M}_{m,k}(\mathcal{E})$, $F \in \mathsf{M}_{k,m}(\mathcal{F})$, such that

$$EE^* \le A + \varepsilon \|X^*X\|^{-1}, \qquad F^*F \le B + \varepsilon \|X^*X\|^{-1}.$$

Then

$$(X^*E)(X^*E)^* \le X^*AX + \varepsilon, \qquad (FX)^*(FX) \le X^*BX + \varepsilon.$$

Since X^*UX is represented by $X^*E \odot FX$, it follows that

$$X^* \begin{pmatrix} A & U \\ U^* & B \end{pmatrix} X = \begin{pmatrix} X^*AX & X^*UX \\ (X^*UX)^* & X^*BX \end{pmatrix} \in \mathsf{M}_n(\mathcal{L})^+,$$

completing the proof.

This last result shows that $\mathcal{L}$ satisfies the hypothesis of Theorem 1.2.7 and so, as an immediate deduction, we have

1.4.5 Corollary. *The space $\mathcal{L}$ is completely order isomorphic to an operator system.*

If $\mathcal{C}$ is any C^*-algebra then the norm and order in a matrix algebra $\mathsf{M}_n(\mathcal{C})$ are linked by $\|C\| \le 1$ in $\mathsf{M}_n(\mathcal{C})$ if and only if

$$\begin{pmatrix} I & \pm C \\ \pm C^* & I \end{pmatrix} \ge 0$$

in $\mathsf{M}_{2n}(\mathcal{C})$. Consequently the complete order isomorphism of the above corollary induces norms on the matrix spaces $\mathsf{M}_n(\mathcal{L})$, $n \ge 1$.

1.4.6 Proposition. *The embeddings* $a \to \begin{pmatrix} a & 0 \\ 0 & 0 \end{pmatrix}$, $b \to \begin{pmatrix} 0 & 0 \\ 0 & b \end{pmatrix}$ *and*

$u \to \begin{pmatrix} 0 & u \\ 0 & 0 \end{pmatrix}$ *of* $\mathcal{A}, \mathcal{B}$ *and* $\mathcal{E} \otimes^h \mathcal{F}$ *respectively into* $\mathcal{L}$ *are complete isometries.*

Proof. The first two are complete order isomorphisms and so are complete isometries. Consider $U \in M_n(\mathcal{E} \otimes^h \mathcal{F})$, $\|U\|_h \leq 1$. Given $\varepsilon > 0$, U has a representation

$$U = E \odot F, \quad E \in M_{n,k}(\mathcal{E}), \quad F \in M_{k,n}(\mathcal{F}),$$

such that $\|E\| < 1 + \varepsilon$, $\|F\| < 1 + \varepsilon$. Then $\begin{pmatrix} 0 & U \\ 0 & 0 \end{pmatrix} \in M_{2n}(\mathcal{E} \otimes^h \mathcal{F})$ has a representation

$$\begin{pmatrix} 0 & U \\ 0 & 0 \end{pmatrix} = E' \odot F'$$

where

$$E' = \begin{pmatrix} E \\ 0 \end{pmatrix} \in M_{2n,k}(\mathcal{E}), \quad F' = (0 \ F) \in M_{k,2n}(\mathcal{F}).$$

Thus

$$E' E'^* = \begin{pmatrix} EE^* & 0 \\ 0 & 0 \end{pmatrix} \leq (1 + \varepsilon)^2 \quad \text{in} \quad M_{2n}(\mathcal{A}),$$

and

$$F'^* F' = \begin{pmatrix} 0 & 0 \\ 0 & F^* F \end{pmatrix} \leq (1 + \varepsilon)^2 \quad \text{in} \quad M_{2n}(\mathcal{B}).$$

Since $\varepsilon > 0$ was arbitrary it follows that

$$\begin{pmatrix} \begin{pmatrix} 1 & 0 \\ 0 & 1 \end{pmatrix} & \pm \begin{pmatrix} 0 & U \\ 0 & 0 \end{pmatrix} \\ \pm \begin{pmatrix} 0 & 0 \\ U^* & 0 \end{pmatrix} & \begin{pmatrix} 1 & 0 \\ 0 & 1 \end{pmatrix} \end{pmatrix} \geq 0 \quad \text{in} \quad M_{2n}(\mathcal{L})$$

and so $\left\| \begin{pmatrix} 0 & U \\ 0 & 0 \end{pmatrix} \right\| \leq 1$.

Conversely, if $\left\| \begin{pmatrix} 0 & U \\ 0 & 0 \end{pmatrix} \right\| \leq 1$ as an element of $M_n(\mathcal{L})$ then

$$\begin{pmatrix} \begin{pmatrix} 1 & 0 \\ 0 & 1 \end{pmatrix} & \begin{pmatrix} 0 & U \\ 0 & 0 \end{pmatrix} \\ \begin{pmatrix} 0 & 0 \\ U^* & 0 \end{pmatrix} & \begin{pmatrix} 1 & 0 \\ 0 & 1 \end{pmatrix} \end{pmatrix} \in M_{2n}(\mathcal{L})^+.$$

Given $\varepsilon > 0$, there exists a representation

$$\begin{pmatrix} 0 & U \\ 0 & 0 \end{pmatrix} = E \odot F, \quad E \in \mathsf{M}_{2n,k}(\mathcal{E}), F \in \mathsf{M}_{k,2n}(\mathcal{F}),$$

such that

$$EE^* \leq 1 + \varepsilon \quad \text{in} \quad \mathsf{M}_{2n}(\mathcal{A}), \qquad F^*F \leq 1 + \varepsilon \quad \text{in} \quad \mathsf{M}_{2n}(\mathcal{B}).$$

Let $X = (1, 0) \in \mathsf{M}_{n,2n}(\mathbf{C})$, $Y = \begin{pmatrix} 0 \\ 1 \end{pmatrix} \in \mathsf{M}_{2n,n}(\mathbf{C})$. Then $(XE) \odot (FY)$ is a representation of $U \in \mathsf{M}_n(\mathcal{E} \otimes^h \mathcal{F})$. Thus

$$\|U\|_h \leq \|XE\| \, \|FY\| \leq 1 + \varepsilon.$$

Since ε was arbitrary, $\|U\|_h \leq 1$, showing that the embedding $U \to \begin{pmatrix} 0 & U \\ 0 & 0 \end{pmatrix}$ is a complete isometry.

1.4.7 Proposition. *If S and T are positive invertible matrices in M_n and $U \in \mathsf{M}_n(\mathcal{E} \otimes^h \mathcal{F})$ then $\begin{pmatrix} S & U \\ U^* & T \end{pmatrix} \in \mathsf{M}_n(\mathcal{L})^+$ if and only if*

$$\|S^{-1/2}UT^{-1/2}\|_h \leq 1.$$

Proof. Since $\begin{pmatrix} S & U \\ U^* & T \end{pmatrix}$ factors as

$$\begin{pmatrix} S^{1/2} & 0 \\ 0 & T^{1/2} \end{pmatrix} \begin{pmatrix} I & S^{-1/2}UT^{-1/2} \\ T^{-1/2}U^*S^{-1/2} & I \end{pmatrix} \begin{pmatrix} S^{1/2} & 0 \\ 0 & T^{1/2} \end{pmatrix},$$

it is clear that positivity of $\begin{pmatrix} S & U \\ U^* & T \end{pmatrix}$ is equivalent to positivity of $\begin{pmatrix} I & V \\ V^* & I \end{pmatrix}$ where $V = S^{-1/2}UT^{-1/2}$. The element $\begin{pmatrix} I & V \\ V^* & I \end{pmatrix}$ is positive if and only if, given $\varepsilon > 0$, there exists a representation $V = E \odot F$ such that $EE^* \leq 1 + \varepsilon$, $F^*F \leq 1 + \varepsilon$. Such conditions can be fulfilled precisely when $\|V\|_h \leq 1$.

We now come to the main results of this section. Recall that, for operator subspaces $\mathcal{E}$ and $\mathcal{F}$ of unital C^*-algebras $\mathcal{A}$ and $\mathcal{B}$, $\mathcal{L}$ is defined to be

$$\left\{ \begin{pmatrix} a & u \\ v & b \end{pmatrix} : a \in \mathcal{A}, b \in \mathcal{B}, u \in \mathcal{E} \otimes \mathcal{F}, v \in \mathcal{F}^* \otimes \mathcal{E}^* \right\}.$$

1.4.8 Theorem. *Let $\mathcal{E} \subseteq \mathcal{A}, \mathcal{F} \subseteq \mathcal{B}$ be operator subspaces of unital C^*-algebras, and let $\phi \colon \mathcal{E} \otimes_h \mathcal{F} \to B(H)$ be a complete contraction. Then there exists a completely positive unital map $\Phi \colon \mathcal{L} \to B(H \oplus H)$ such that*

$$\Phi \begin{pmatrix} \lambda & u \\ v^* & \mu \end{pmatrix} = \begin{pmatrix} \lambda & \phi(u) \\ \phi(v)^* & \mu \end{pmatrix}$$

for $\lambda, \mu \in \mathbb{C}$, $u, v \in \mathcal{E} \otimes^h \mathcal{F}$.

Proof. Let $\mathcal{L}_1$ be the subspace

$$\left\{ \begin{pmatrix} \lambda & u \\ v^* & \mu \end{pmatrix} : \lambda, \mu \in \mathbb{C}, u, v \in \mathcal{E} \otimes^h \mathcal{F} \right\} \subseteq \mathcal{L}$$

and define $\psi \colon \mathcal{L}_1 \to B(H \oplus H)$ by

$$\psi \begin{pmatrix} \lambda & u \\ v^* & \mu \end{pmatrix} = \begin{pmatrix} \lambda & \phi(u) \\ \phi(v)^* & \mu \end{pmatrix}.$$

Consider $\begin{pmatrix} S & U \\ U^* & T \end{pmatrix} \in M_n(\mathcal{L}_1)^+$, and suppose initially that S and T are invertible. By Proposition 1.4.7,

$$\| S^{-1/2} U T^{-1/2} \|_h \le 1,$$

so

$$\| S^{-1/2} \phi_n(U) T^{-1/2} \| \le 1,$$

using $\|\phi\|_{cb} \le 1$. Then another application of Proposition 1.4.7 shows that $\begin{pmatrix} S & \phi_n(U) \\ \phi_n(U)^* & T \end{pmatrix} \ge 0$. In the general case we add ε to S and T to ensure invertibility, apply the previous reasoning to obtain $\begin{pmatrix} S+\varepsilon & \phi_n(U) \\ \phi_n(U)^* & T+\varepsilon \end{pmatrix} \ge 0$ and then let ε tend to 0. This shows that $\psi \colon \mathcal{L}_1 \to B(H \oplus H)$ is completely positive, and it is unital by construction. Now Φ is obtained by applying Arveson's Hahn–Banach theorem (1.2.3) to ψ.

1.4.9 Theorem. *Let $\phi \colon \mathcal{E} \otimes_h \mathcal{F} \to B(H)$ be a complete contraction and let $\Phi \colon \mathcal{L} \to B(H \oplus H)$ be the unital completely positive map constructed in Theorem 1.4.8. Then there exist unital completely positive maps $\psi_1 \colon \mathcal{A} \to B(H)$, $\psi_2 \colon \mathcal{B} \to B(H)$ such that*

$$\Phi \begin{pmatrix} a & u \\ v^* & b \end{pmatrix} = \begin{pmatrix} \psi_1(a) & \phi(u) \\ \phi(v)^* & \psi_2(b) \end{pmatrix}$$

for $a \in \mathcal{A}, b \in \mathcal{B}, u, v \in \mathcal{E} \otimes^h \mathcal{F}$.

Proof. For every self-adjoint element $a \in \mathcal{A}$

$$\begin{pmatrix} -\|a\| & 0 \\ 0 & 0 \end{pmatrix} \le \begin{pmatrix} a & 0 \\ 0 & 0 \end{pmatrix} \le \begin{pmatrix} \|a\| & 0 \\ 0 & 0 \end{pmatrix}$$

and so

$$\begin{pmatrix} -\|a\| & 0 \\ 0 & 0 \end{pmatrix} \leq \Phi \begin{pmatrix} a & 0 \\ 0 & 0 \end{pmatrix} \leq \begin{pmatrix} \|a\| & 0 \\ 0 & 0 \end{pmatrix}.$$

Then $\Phi \begin{pmatrix} a & 0 \\ 0 & 0 \end{pmatrix}$ can only have a non-zero entry in the (1,1) position. Thus there exists a completely positive map $\psi_1 \colon \mathcal{A} \to B(H)$ such that $\begin{pmatrix} \psi_1(a) & 0 \\ 0 & 0 \end{pmatrix} = \Phi \begin{pmatrix} a & 0 \\ 0 & 0 \end{pmatrix}$. The same argument applies to the (2,2) position and so Φ has the desired form.

Our discussion of the Haagerup tensor product is far from complete, but in these notes our objective is to obtain the representation theorems for multilinear maps as quickly as possible. These last two theorems provide the required link between the single and multivariable theories.

1.5 Multilinear Maps

Cohomology theory requires the study of multilinear maps and so we now discuss the extension of the linear results to this more general setting. The main difficulty is to pass from the linear to the bilinear case. Once this has been achieved, the multilinear results are obtained inductively.

Let $\mathcal{E}$ and $\mathcal{F}$ be subspaces of unital C^*-algebras $\mathcal{A}$ and $\mathcal{B}$ respectively. A bilinear map $\phi \colon \mathcal{E} \times \mathcal{F} \to B(H)$ is bounded if

$$\sup\{\|\phi(e,f)\| \colon \|e\|, \|f\| \leq 1\}$$

is finite, in which case $\|\phi\|$ denotes the supremum. There is also a notion of complete boundedness for bilinear maps which we now introduce. For each $n \geq 1$, define $\phi_n \colon \mathsf{M}_n(\mathcal{E}) \times \mathsf{M}_n(\mathcal{F}) \to \mathsf{M}_n(B(H))$ by specifying the (i,j) entry of $\phi_n((e_{ij}), (f_{ij}))$ to be $\sum_k \phi(e_{ik}, f_{kj})$. It is helpful to think of this as matrix multiplication, but with ϕ replacing the product. We say that ϕ is *completely bounded* if $\sup_{n \geq 1} \|\phi_n\|$ is finite, in which case this quantity is denoted $\|\phi\|_{cb}$. The "matrix multiplication" definition of ϕ_n implies that if $U, V, W \in \mathsf{M}_n(\mathbb{C})$, $E \in \mathsf{M}_n(\mathcal{E})$, $F \in \mathsf{M}_n(\mathcal{E})$, then

$$U\phi_n(EW, F)V = \phi_n(UE, WFV).$$

This will be used subsequently several times. If $\pi \colon \mathcal{A} \to B(K_1)$, $\rho \colon \mathcal{B} \to B(K_2)$ are representations and $W \colon H \to K_2$, $T \colon K_2 \to K_1$, $V \colon K_1 \to H$ are bounded operators, then $\phi \colon \mathcal{E} \times \mathcal{F} \to B(H)$, defined by

$$\phi(e,f) = V\pi(e)T\rho(f)W$$

is easily checked to be completely bounded with $\|\phi\|_{cb} \leq \|V\|\,\|T\|\,\|W\|$. One of the main results of this section states that every completely bounded

bilinear map has such a form (compare with Theorem 1.3.1). The order of
the variables is important because of the nature of matrix multiplication.
In general the map $(x, y) \to yx$ will not be completely bounded.

With each bilinear map $\phi \colon \mathcal{E} \times \mathcal{F} \to B(H)$ we may associate a linear map
$\tilde{\phi} \colon \mathcal{E} \otimes^h \mathcal{F} \to B(H)$ on the (uncompleted) Haagerup tensor product by

$$\tilde{\phi}(e \otimes f) = \phi(e, f), \qquad e \in \mathcal{E}, \quad f \in \mathcal{F}.$$

The connection between complete boundedness and the Haagerup tensor
product is exhibited by the next result.

1.5.1 Proposition. *A bilinear map $\phi \colon \mathcal{E} \times \mathcal{F} \to B(H)$ is completely
bounded if and only if $\tilde{\phi} \colon \mathcal{E} \otimes^h \mathcal{F} \to B(H)$ is completely bounded, in which
case $\tilde{\phi}$ extends uniquely to a completely bounded map on $\mathcal{E} \otimes_h \mathcal{F}$, and
satisfies $\|\tilde{\phi}\|_{cb} = \|\phi\|_{cb}$.*

Proof. Consider $U \in M_n(\mathcal{E} \otimes^h \mathcal{F})$, $\|U\|_h < 1$. Then there is a representation $U = E \odot F$ with $E \in M_{n,k}(\mathcal{E})$, $F \in M_{k,n}(\mathcal{F})$ and $\|E\|, \|F\| < 1$.
Let $r = \max\{n, k\}$. By adding rows and columns of zeros to E and F we
obtain new matrices $E_1 \in M_r(\mathcal{E})$, $F_1 \in M_r(\mathcal{F})$ such that $\tilde{\phi}_n(U)$ is the upper
left-hand $n \times n$ corner of $\phi_r(E_1, F_1)$. Then, if ϕ is completely bounded,

$$\|\tilde{\phi}_n(U)\| \leq \|\phi_r(E_1, F_1)\| \leq \|\phi\|_{cb},$$

showing that $\|\tilde{\phi}_n\| \leq \|\phi\|_{cb}$. Since n was arbitrary, this gives $\|\tilde{\phi}\|_{cb} \leq \|\phi\|_{cb}$.

Conversely suppose that $\tilde{\phi}$ is completely bounded, and consider $E \in
M_n(\mathcal{E})$, $F \in M_n(\mathcal{F})$ with $\|E\|, \|F\| \leq 1$. Then set $U = E \odot F \in M_n(\mathcal{E} \otimes^h \mathcal{F})$.
By definition, $\|U\|_h \leq 1$ so

$$\|\phi_n(E, F)\| = \|\tilde{\phi}_n(U)\|_h \leq \|\tilde{\phi}\|_{cb}.$$

Thus $\|\phi_n\| \leq \|\tilde{\phi}\|_{cb}$. Again, n was arbitrary, and so $\|\phi\|_{cb} \leq \|\tilde{\phi}\|_{cb}$, establishing equality.

The extension of $\tilde{\phi}$ to $\mathcal{E} \otimes_h \mathcal{F}$ is clear. $\quad\blacksquare$

In general complete boundedness of ϕ is not equivalent to boundedness
of $\tilde{\phi}$, although we will encounter some special circumstances in which this is
true. We present one now, in which we assume that H is one dimensional.
Of course, this means that $B(H)$ is identified with $\mathbb{C}$.

1.5.2 Theorem. *The bilinear form $\phi \colon \mathcal{E} \times \mathcal{F} \to \mathbb{C}$ is completely bounded
if and only if $\tilde{\phi} \colon \mathcal{E} \otimes_h \mathcal{F} \to \mathbb{C}$ is bounded, and $\|\tilde{\phi}\| = \|\phi\|_{cb}$.*

Proof. If ϕ is completely bounded then, by Proposition 1.5.1, $\tilde{\phi}$ is completely bounded and so is bounded. Moreover

$$\|\phi\|_{cb} = \|\tilde{\phi}\|_{cb} \geq \|\tilde{\phi}\|.$$

Conversely, suppose that $\tilde{\phi}$ is bounded. Fix $n \geq 1$, and consider $E \in M_n(\mathcal{E})$, $F \in M_n(\mathcal{F})$ of unit norm. Then there exist unit vectors $\xi, \eta \in \mathbf{C}^n$ such that

$$\|\phi_n(E, F)\| = \langle \phi_n(E, F)\xi, \eta \rangle.$$

Let ξ_1 be the first standard basis vector of $\mathbf{C}^n$ and choose unitary matrices $U, V \in M_n$ such that $U\xi_1 = \xi$, $V\xi_1 = \eta$. Then

$$\|\phi_n(E, F)\| = \langle V^*\phi_n(E, F)U\xi_1, \xi_1 \rangle$$
$$= \langle \phi_n(V^*E, FU)\xi_1, \xi_1 \rangle.$$

Let $E_1 \in M_{1,n}(\mathcal{E})$ be the first row of V^*E, let $F_1 \in M_{n,1}(\mathcal{F})$ be the first column of FU and let $u = E_1 \odot F_1 \in \mathcal{E} \otimes_h \mathcal{F}$. Then $\|u\|_h \leq 1$ and

$$\|\phi_n(E, F)\| = \tilde{\phi}(u) \leq \|\tilde{\phi}\|.$$

Thus $\|\phi_n\| \leq \|\tilde{\phi}\|$ for all $n \geq 1$, which shows that ϕ is completely bounded and $\|\phi\|_{cb} \leq \|\tilde{\phi}\|$. We have now proved equality of the two norms.

This result identifies the dual of $\mathcal{E} \otimes_h \mathcal{F}$ as the space of completely bounded bilinear functionals on $\mathcal{E} \times \mathcal{F}$.

We now come to the representation theorem for completely bounded bilinear maps. We will need one technical result.

1.5.3 Lemma. *Suppose that* $\begin{pmatrix} X & Z \\ Z^* & Y \end{pmatrix} \in M_2(B(H))^+$. *If* $h, k \in H$ *then*

$$|\langle Zk, h \rangle|^2 \leq \langle Xh, h \rangle \langle Yk, k \rangle.$$

Proof. Let t and θ be arbitrary real numbers. Then

$$\left\langle \begin{pmatrix} X & Z \\ Z^* & Y \end{pmatrix} \begin{pmatrix} te^{i\theta}h \\ k \end{pmatrix}, \begin{pmatrix} te^{i\theta}h \\ k \end{pmatrix} \right\rangle \geq 0,$$

which is equivalent to

$$t^2\langle Xh, h \rangle + te^{i\theta}\langle Z^*h, k \rangle + te^{-i\theta}\langle Zk, h \rangle + \langle Yk, k \rangle \geq 0.$$

An appropriate choice of θ reduces this to

$$t^2\langle Xh, h \rangle + 2t|\langle Zk, h \rangle| + \langle Yk, k \rangle \geq 0$$

for $t \in \mathbf{R}$, and the result follows.

1.5.4 Theorem. *Let $\mathcal{E}$ and $\mathcal{F}$ be operator subspaces of C^*-algebras $\mathcal{A}$ and $\mathcal{B}$. If $\phi\colon \mathcal{E} \times \mathcal{F} \to B(H)$ is completely bounded then there exist representations $\pi\colon \mathcal{A} \to B(K_1)$, $\rho\colon \mathcal{B} \to B(K_2)$ and bounded operators $W\colon H \to K_2$, $T\colon K_2 \to K_1$, $V\colon K_1 \to H$ satisfying $\|V\|\,\|T\|\,\|W\| = \|\phi\|_{cb}$ such that*

$$\phi(e, f) = V\pi(e)T\rho(f)W, \qquad e \in \mathcal{E}, \quad f \in \mathcal{F}.$$

Proof. Scaling allows us to assume that $\|\phi\|_{cb} = 1$ and so the associated map $\tilde{\phi}\colon \mathcal{E} \otimes_h \mathcal{F} \to B(H)$ is also completely bounded with $\|\tilde{\phi}\|_{cb} = 1$, by Proposition 1.5.1.

From Theorem 1.4.9 there exist completely positive unital maps $\psi_1\colon \mathcal{A} \to B(H)$, $\psi_2\colon \mathcal{B} \to B(H)$ such that the map

$$\begin{pmatrix} a & u \\ v^* & b \end{pmatrix} \to \begin{pmatrix} \psi_1(a) & \tilde{\phi}(u) \\ \tilde{\phi}(v)^* & \psi_2(b) \end{pmatrix}$$

is completely positive on $\mathcal{L}$. By Theorem 1.2.1 there exist representations $\pi\colon \mathcal{A} \to B(K_1)$, $\rho\colon \mathcal{B} \to B(K_2)$ and contractions $V_1\colon H \to K_1$, $V_2\colon H \to K_2$ such that

$$\psi_1(a) = V_1^*\pi(a)V_1, \quad \psi_2(b) = V_2^*\rho(b)V_2, \qquad a \in \mathcal{A}, \quad b \in \mathcal{B}.$$

Consider elements $e_i \in \mathcal{E}, f_i \in \mathcal{F}, h_i, k_i \in H, 1 \leq i \leq n$, and let $E = \begin{pmatrix} e_1 \\ \vdots \\ e_n \end{pmatrix} \in \mathsf{M}_{n,1}(\mathcal{E}), F = (f_1, \ldots, f_n) \in \mathsf{M}_{1,n}(\mathcal{F}), U = E \odot F \in \mathsf{M}_n(\mathcal{E} \otimes^h \mathcal{F})$.
Then

$$\begin{pmatrix} EE^* & U \\ U^* & F^*F \end{pmatrix} \in \mathsf{M}_n(\mathcal{L})^+$$

so

$$\begin{pmatrix} (\psi_1)_n(EE^*) & \tilde{\phi}_n(U) \\ \tilde{\phi}_n(U)^* & (\psi_2)_n(F^*F) \end{pmatrix} \geq 0.$$

By Lemma 1.5.3

$$|\langle \tilde{\phi}_n(U)k, h\rangle|^2 \leq \langle (\psi_1)_n(EE^*)h, h\rangle \langle (\psi_2)_n(F^*F)k, k\rangle$$

where $h = (h_1, \ldots, h_n), k = (k_1, \ldots, k_n) \in H \oplus \cdots \oplus H$. Thus

$$\left| \sum_{i,j} \langle \tilde{\phi}(e_i \otimes f_j)k_j, h_i\rangle \right|^2$$

$$\leq \left(\sum_{i,j} \langle V_1^*\pi(e_ie_j^*)V_1h_j, h_i\rangle \right) \left(\sum_{i,j} \langle V_2^*\rho(f_i^*f_j)V_2k_j, k_i\rangle \right)$$

$$= \left\| \sum_i \pi(e_i^*)V_1h_i \right\|^2 \left\| \sum_i \rho(f_i)V_2k_i \right\|^2.$$

Let J_1 and J_2 be respectively the spans of

$$\{\pi(e^*)V_1 h\colon\ e \in \mathcal{E}, h \in H\} \quad \text{and} \quad \{\rho(f)V_2 k\colon\ f \in \mathcal{F}, k \in H\}.$$

The previous inequality shows that there is a well–defined sesquilinear form, of norm at most 1, on $J_2 \times J_1$ given by

$$\left[\sum_j \rho(f_j)V_2 k_j, \sum_i \pi(e_i^*)V_1 h_i\right] = \sum_{i,j}\langle \tilde{\phi}(e_i \otimes f_j)k_j, h_i\rangle.$$

This extends to a contractive sesquilinear form on $K_2 \times K_1$ by first passing to the completions of J_1 and J_2 and then letting it be 0 on the orthogonal complements. Then there exists $T\colon K_2 \to K_1$, $\|T\| \leq 1$, such that

$$\begin{aligned}
\langle \tilde{\phi}(e \otimes f)k, h\rangle &= [\rho(f)V_2 k, \pi(e^*)V_1 h]\\
&= \langle T\rho(f)V_2 k, \pi(e^*)V_1 h\rangle\\
&= \langle V_1^*\pi(e)T\rho(f)V_2 k, h\rangle.
\end{aligned}$$

Thus $\phi(e, f) = V\pi(e)T\rho(f)W$, where $V = V_1^*$ and $W = V_2$.

We have $\|V\|\,\|T\|\,\|W\| \leq 1$ since each operator is a contraction. Equality must hold since otherwise we could easily estimate that $\|\phi\|_{cb} < 1$.

This theorem contains a Hahn–Banach theorem for completely bounded bilinear maps.

1.5.5 Theorem. *Let $\mathcal{E}_1 \subseteq \mathcal{E}_2 \subseteq A$ and $\mathcal{F}_1 \subseteq \mathcal{F}_2 \subseteq B$ be operator subspaces of C^*-algebras A and B. If $\phi\colon \mathcal{E}_1 \times \mathcal{F}_1 \to B(H)$ is completely bounded then there exists an extension $\psi\colon \mathcal{E}_2 \times \mathcal{F}_2 \to B(H)$ such that $\|\psi\|_{cb} = \|\phi\|_{cb}$.*

Proof. Let $\phi(e, f) = V\pi(e)T\rho(f)W$ ($e \in \mathcal{E}_1, f \in \mathcal{F}_1$) be the representation obtained in Theorem 1.5.4. The extension ψ is defined by

$$\psi(e, f) = V\pi(e)T\rho(f)W, \qquad e \in \mathcal{E}_2, \quad f \in \mathcal{F}_2.$$

In order to obtain the multilinear version of Theorem 1.5.4 we must define the Haagerup norm on the tensor product of k operator spaces. If $\mathcal{E}_1, \ldots, \mathcal{E}_k$ are operator spaces and

$$E_1 \in \mathsf{M}_{n,n_1}(\mathcal{E}_1), E_i \in \mathsf{M}_{n_{i-1},n_i}(\mathcal{E}), \quad 2 \leq i \leq k-1, \ E_k \in \mathsf{M}_{n_{k-1},n}(\mathcal{E}_k)$$

then we define $U \in \mathsf{M}_n(\mathcal{E}_1 \otimes \cdots \otimes \mathcal{E}_k)$ by first forming $E_1 \odot E_2$, then $(E_1 \odot E_2) \odot E_3$ and so on. Then we define

$$\|U\|_h = \inf\{\|E_1\|\,\|E_2\| \ldots \|E_k\|\colon\ U = E_1 \odot E_2 \ldots \odot E_k\}.$$

Corollary 1.4.5 and Proposition 1.4.6 show that the Haagerup tensor product of two operator spaces is again an operator space. It is then clear from the definition of the norm that we have associativity:

$$(\mathcal{E}_1 \otimes_h \mathcal{E}_2) \otimes_h \mathcal{E}_3 = \mathcal{E}_1 \otimes_h (\mathcal{E}_2 \otimes_h \mathcal{E}_3) = \mathcal{E}_1 \otimes_h \mathcal{E}_2 \otimes_h \mathcal{E}_3.$$

If $\phi\colon \mathcal{E}_1 \times \cdots \times \mathcal{E}_k \to B(H)$ is a k-linear map then

$$\phi_n\colon \mathsf{M}_n(\mathcal{E}_1) \times \cdots \times \mathsf{M}_n(\mathcal{E}_k) \to \mathsf{M}_n(B(H))$$

is defined analogously to the bilinear case: mimic matrix multiplication but replace the product by the k-linear map. Then $\|\phi\|_{cb} = \sup_{n \geq 1} \|\phi_n\|$ if this supremum is finite. There is a linearization $\tilde{\phi}\colon \mathcal{E}_1 \otimes_h \cdots \otimes_h \mathcal{E}_k \to B(H)$ of ϕ and, following the proof of Proposition 1.5.1, $\|\phi\|_{cb} = \|\tilde{\phi}\|_{cb}$.

1.5.6 Theorem. *Let $\mathcal{E}_i \subseteq \mathcal{A}_i$, $1 \leq i \leq k$, be operator subspaces of C^*-algebras and let $\phi\colon \mathcal{E}_1 \times \cdots \times \mathcal{E}_k \to B(H)$ be completely bounded. Then there exist representations $\pi_i\colon \mathcal{A}_i \to B(K_i)$, $1 \leq i \leq k$, and bounded operators $V_1\colon K_1 \to H$, $V_{k+1}\colon H \to K_k$, $V_i\colon K_i \to K_{i-1}$, $2 \leq i \leq k$, such that*

$$\phi(e_1, e_2, \ldots, e_k) = V_1 \pi_1(e_1) V_2 \pi_2(e_2) \ldots V_k \pi_k(e_k) V_{k+1}$$

and $\|V_1\| \|V_2\| \ldots \|V_k\| = \|\phi\|_{cb}$.

Proof. The proof is by induction on k. The step from $k = 2$ to $k = 3$ contains all the ingredients of the general argument and so we give only the simpler case. We assume without loss of generality that $\|\phi\|_{cb} = 1$.

Now $\mathcal{E}_2 \otimes_h \mathcal{E}_3$ is an operator subspace of a C^*-algebra $\mathcal{C}$. Thus

$$\tilde{\phi}\colon \mathcal{E}_1 \otimes_h \mathcal{E}_2 \otimes_h \mathcal{E}_3 \to B(H)$$

may be viewed as a map $\tilde{\phi}\colon \mathcal{E}_1 \otimes_h \mathcal{F} \to B(H)$ where $\mathcal{F} = \mathcal{E}_2 \otimes_h \mathcal{E}_3$. We apply the bilinear case (Theorem 1.5.4). Then $\tilde{\phi}$ has the form

$$\tilde{\phi}(e_1 \otimes f) = V_1 \pi_1(e_1) T \rho(f) W$$

where ρ is a representation of $\mathcal{C}$ and $\|V_1\|, \|T\|, \|W\| \leq 1$.

The map $f \to \rho(f)$ is a complete contraction on $\mathcal{E}_2 \otimes_h \mathcal{E}_3$, so has the form

$$\rho(e_2 \otimes e_3) = \tilde{V}_2 \pi_2(e_2) V_3 \pi_3(e_3) \tilde{V}_4,$$

$\|\tilde{V}_2\|, \|V_3\|, \|\tilde{V}_4\| \leq 1$. Then

$$\phi(e_1, e_2, e_3) = V_1 \pi_1(e_1) V_2 \pi_2(e_2) V_3 \pi_3(e_3) V_4$$

where $V_2 = T\tilde{V}_2$ and $V_4 = \tilde{V}_4 W$. It is clear that $\|V_i\| \leq 1$ for $1 \leq i \leq 4$.

We mention in passing that there is also a Hahn–Banach theorem in the multilinear case. This is identical to Theorem 1.5.5 and so we omit it.

For a von Neumann algebra $\mathcal{M} \subseteq B(K)$, recall that normal representations $\pi \colon \mathcal{M} \to B(H)$ have a particularly simple structure. For an index set Λ, let K_Λ denote $\bigoplus_{\lambda \in \Lambda} K_\lambda$, where each K_λ is a copy of K, and let $\rho \colon \mathcal{M} \to B(K_\Lambda)$ be the normal representation where $\rho(m)$ acts as m on each copy of K (called an amplification). Then there exists a projection $P \in \rho(\mathcal{M})'$ (the commutant) and a unitary $U \colon H \to PK_\Lambda$ such that

$$\pi(m) = U^*\rho(m)PU, \qquad m \in \mathcal{M},$$

[Di2, 1.4.4]. Then a map $V_1^*\pi V_2$ becomes $(V_1^*U^*)\rho(PUV_2)$ which has the form $W_1^*\rho W_2$, where ρ is an amplification of the identity representation of a suitable cardinality. Any completely bounded map with such a representation is normal and we now address the converse problem, which is of crucial importance for investigating cohomology. We consider the linear case first and then deduce the multilinear result.

1.5.7 Lemma. *Let $\mathcal{M}$ be a von Neumann algebra and let $\phi \colon \mathcal{M} \to B(H)$ be a completely bounded normal map.*

(a) If ϕ has a representation $V_1^*\pi V_2$ where $V_i \colon H \to K$, $i = 1,2$, and $\pi \colon \mathcal{M} \to B(K)$, then there is a central projection $z \in \pi(\mathcal{M})''$ (depending only on π) such that

$$\phi(m) = V_1^*\pi(m)zV_2, \qquad m \in \mathcal{M},$$

and $m \to \pi(m)z$ is a normal representation of $\mathcal{M}$,
(b) Further, ϕ has a representation $W_1^*\rho W_2$ where ρ is an amplification of the identity.

Proof. The remarks preceding this result show that (b) is a consequence of (a), so we only prove the first part.

Following [T, III.2.14], there is a central projection $z \in \pi(\mathcal{M})''$ such that πz is normal and the functionals $m \to \langle \pi(m)\xi_2, \xi_1 \rangle$ are normal (respectively singular) if $\xi_1, \xi_2 \in zK$ (respectively $\xi_1, \xi_2 \in (1-z)K$). If $\eta_1, \eta_2 \in H$, we let $\xi_i = V_i\eta_i \in K$, $i = 1,2$. Then

$$\langle \phi(m)\eta_2, \eta_1 \rangle = \langle \pi(m)zV_2\eta_2, zV_1\eta_1 \rangle + \langle \pi(m)(1-z)V_2\eta_2, (1-z)V_1\eta_1 \rangle$$

for $m \in \mathcal{M}$, since z and $(1-z)$ commute with $\pi(\mathcal{M})$. Thus

$$\langle \phi(m)\eta_2, \eta_1 \rangle - \langle \pi(m)z\xi_2, z\xi_1 \rangle = \langle \pi(m)(1-z)\xi_2, (1-z)\xi_1 \rangle$$

and so the left–hand side defines a normal functional while the right–hand side is singular. Thus $\langle \pi(m)(1-z)\xi_2, (1-z)\xi_1 \rangle = 0$, and

$$\langle V_1^*\pi(m)(1-z)V_2\eta_2, \eta_1 \rangle = 0, \qquad m \in \mathcal{M}, \quad \eta_1, \eta_2 \in H.$$

It follows that $\phi = V_1^*\pi zV_2$ where πz is a normal representation of $\mathcal{M}$.

1.5.8 Theorem. *Let ϕ: $\mathcal{M} \times \cdots \times \mathcal{M} \to B(H)$ be an n-linear completely bounded map which is separately normal in each variable. Then ϕ may be represented by*

$$\phi = V_1 \rho_1 V_2 \ldots V_n \rho_n V_{n+1}$$

where each ρ_i is an amplification of the identity representation.

Proof. By Theorem 1.5.6, ϕ has a representation

$$\phi(m_1, m_2, \ldots, m_n) = W_1 \pi_1(m_1) W_2 \pi_2(m_2) \ldots W_n \pi_n(m_n) W_{n+1}, \quad m_i \in \mathcal{M}.$$

We fix $m_2, \ldots, m_n$ to obtain a normal linear completely bounded map

$$\psi_1(m) = \phi(m, m_2, \ldots, m_n) = W_1 \pi_1(m)(W_2 \pi_2(m_2) \ldots W_n \pi_n(m_n) W_{n+1})$$

and so, by Lemma 1.5.7, there exists $z_1 \in \pi_1(\mathcal{M})''$ such that $\pi_1 z_1$ is a normal representation and

$$\phi(m_1, m_2, \ldots, m_n) = W_1 \pi_1(m_1) z_1 W_2 \pi_2(m_2) \ldots W_n \pi_n(m_n) W_{n+1}.$$

We then repeat this argument inductively to replace each π_i by $\pi_i z_i$ for z_i in the centre of $\pi_i(\mathcal{M})''$, so that $\pi_i z_i$ is a normal representation and ϕ is still represented by the product after the insertion of these central projections. At the kth stage, apply Lemma 1.5.7 to the normal linear completely bounded map obtained by fixing $m_1, \ldots, m_{k-1}, m_{k+1}, \ldots, m_n$.

Each normal representation $\pi_i z_i$ has the form $U_i^* \rho_i P_i U_i$ for an amplification ρ_i of the identity, a projection $P_i \in \rho_i(\mathcal{M})'$ and a unitary U_i, so

$$\phi(m_1, \ldots, m_n) = (W_1 U_1^*) \rho_1(m_1)(P_1 U_1 W_2 U_2^*) \rho_2(m_2) \ldots \rho_n(m_n)(P_n U_n W_{n+1})$$

which is of the desired form.

1.6 Automatic Complete Boundedness

The representation theorems for completely bounded linear and multilinear maps (1.3.1 and 1.5.6) provide powerful tools for calculating cohomology groups. Unfortunately, there are many examples of maps which fail to be completely bounded, and there is no useful representation for a general linear or multilinear map. The usefulness of completely bounded maps may be attributed to the fact that the standard maps of operator algebras, such as representations, are all completely bounded. We cannot expect any simple combination of such maps to represent a map which is not completely bounded. Instead we look for conditions which will imply complete boundedness, and which arise naturally in cohomology theory. These criteria are met by modularity with respect to C^*-subalgebras.

Let $\mathcal{A}$ and $\mathcal{B}$ be C^*-subalgebras of $B(H)$ and let $\mathcal{E}$ be an operator subspace of $B(H)$ which is also a module with respect to operator multiplication on the left by $\mathcal{A}$ and on the right by $\mathcal{B}$. We say that $\mathcal{E}$ is an $(\mathcal{A}, \mathcal{B})$-bimodule. Any operator subspace is a $(\mathbb{C}, \mathbb{C})$-bimodule and so there is no loss of generality in assuming that $\mathcal{A}$ and $\mathcal{B}$ are unital. It is not necessary for $\mathcal{E}$ to contain $\mathcal{A}$ and $\mathcal{B}$; for example, let $\mathcal{E} = K(H)$, $\mathcal{A} = \mathcal{B} = B(H)$. A map $\phi \colon \mathcal{E} \to B(H)$ is said to be an $(\mathcal{A}, \mathcal{B})$-*bimodule map* if

$$\phi(aeb) = a\phi(e)b, \qquad a \in \mathcal{A}, b \in \mathcal{B}, e \in \mathcal{E}.$$

In general nothing further can be said, because all linear maps are $(\mathbb{C}, \mathbb{C})$-bimodular. However, if $\mathcal{A}$ and $\mathcal{B}$ are large in a suitable sense then complete boundedness is automatic. For a fixed vector $\xi \in H$, $\overline{\mathcal{A}\xi}$ will denote the norm closed cyclic subspace generated by $\{a\xi \colon a \in \mathcal{A}\}$. We say that ξ is a *cyclic vector* (for $\mathcal{A}$) if $\overline{\mathcal{A}\xi} = H$.

1.6.1 Theorem. *Let $\mathcal{A}$ and $\mathcal{B}$ be unital C^*-subalgebras of $B(H)$, let $\mathcal{E} \subseteq B(H)$ be an $(\mathcal{A}, \mathcal{B})$-bimodule and let $\phi \colon \mathcal{E} \to B(H)$ be a bounded $(\mathcal{A}, \mathcal{B})$-bimodule map. Further suppose that $\mathcal{A}$ and $\mathcal{B}$ satisfy one of the following hypotheses:*

(i) $\mathcal{A}$ and $\mathcal{B}$ have cyclic vectors,

(ii) given any finite collection of vectors ξ_i, $\eta_j \in H$, $1 \leq i, j \leq n$, there exist $\xi, \eta \in H$ such that $\xi_i \in \overline{\mathcal{A}\xi}$, $\eta_j \in \overline{\mathcal{B}\eta}$, $1 \leq i, j \leq n$.

Then ϕ is completely bounded and $\|\phi\|_{cb} = \|\phi\|$.

Proof. Since (ii) is implied by (i) we will assume that the second hypothesis is satisfied. The inequality $\|\phi_n\| \geq \|\phi\|$ is always true, where $\phi_n \colon \mathsf{M}_n(\mathcal{E}) \to \mathsf{M}_n(B(H))$, so it suffices to prove that $\|\phi_n\| \leq \|\phi\|$ for $n \geq 1$. We argue by contradiction, so suppose that $\|\phi_n\| > \|\phi\|$ for some fixed integer n. Without loss of generality we take $\|\phi\|$ to be 1. Then there exists $(e_{ij}) \in \mathsf{M}_n(\mathcal{E})$, $\|(e_{ij})\| \leq 1$, and vectors $\xi_i, \eta_j \in H$, $1 \leq i, j \leq n$ such that

$$\sum_{i=1}^n \|\xi_i\|^2, \quad \sum_{j=1}^n \|\eta_j\|^2 < 1$$

and

$$\left| \sum_{i,j=1}^n \langle \phi(e_{ij})\eta_j, \xi_i \rangle \right| > 1.$$

By hypothesis, choose $\xi, \eta \in H$ such that $\xi_i \in \overline{\mathcal{A}\xi}$, $\eta_j \in \overline{\mathcal{B}\eta}$, $1 \leq i, j \leq n$. Then we may find elements $a_i \in \mathcal{A}$, $b_j \in \mathcal{B}$ such that

$$\sum_{i=1}^n \|a_i\xi\|^2, \quad \sum_{j=1}^n \|b_j\eta\|^2 < 1$$

and

$$\left| \sum_{i,j=1}^{n} \langle \phi(e_{ij}) b_j \eta, a_i \xi \rangle \right| > 1.$$

Fix an arbitrary $\varepsilon > 0$ and write

$$a = \varepsilon + \sum_{i=1}^{n} a_i^* a_i, \quad b = \varepsilon + \sum_{j=1}^{n} b_j^* b_j,$$

noting that a and b are invertible positive elements. Now define $\tilde{\xi} = a^{1/2}\xi$, $\tilde{\eta} = b^{1/2}\eta$, $c_i = a_i a^{-1/2}$ and $d_j = b_j b^{-1/2}$. Then

$$\sum_{i=1}^{n} c_i^* c_i = \sum_{i=1}^{n} a^{-1/2} a_i^* a_i a^{-1/2} \le a^{-1/2} \left(\varepsilon + \sum_{i=1}^{n} a_i^* a_i \right) a^{-1/2} = 1,$$

and similarly $\sum_{i=1}^{n} d_j^* d_j \le 1$. Also

$$\|\tilde{\xi}\|^2 = \langle a^{1/2}\xi, a^{1/2}\xi \rangle = \langle a\xi, \xi \rangle$$

$$= \varepsilon \|\xi\|^2 + \sum_{i=1}^{n} \langle a_i^* a_i \xi, \xi \rangle$$

$$= \varepsilon \|\xi\|^2 + \sum_{i=1}^{n} \|a_i \xi\|^2$$

and the same calculation shows that $\|\tilde{\eta}\|^2 = \varepsilon \|\eta\|^2 + \sum_{j=1}^{n} \|b_j \eta\|^2$. By construction

$$c_i \tilde{\xi} = a_i \xi, \quad d_j \tilde{\eta} = b_j \eta, \quad 1 \le i, j \le n,$$

and so

$$\left| \sum_{i,j=1}^{n} \langle c_i^* \phi(e_{ij}) d_j \tilde{\eta}, \tilde{\xi} \rangle \right| = \left| \sum_{i,j=1}^{n} \langle \phi(e_{ij}) b_j \eta, a_i \xi \rangle \right| > 1.$$

Let $e \in \mathcal{E}$ be defined by the matrix product

$$e = (c_1^*, \dots, c_n^*)(e_{ij}) \begin{pmatrix} d_1 \\ \vdots \\ d_n \end{pmatrix}.$$

Since $\sum_{i=1}^{n} c_i^* c_i, \sum_{j=1}^{n} d_j^* d_j \le 1$, it is clear that $\|e\| \le 1$. Using the bimodular properties of ϕ, we obtain

$$|\langle \phi(e)\tilde{\eta}, \tilde{\xi} \rangle| > 1.$$

If we now choose ε to be suitably small, then our previous inequalities show that $\|\tilde{\xi}\|, \|\tilde{\eta}\| < 1$, and we have contradicted $\|\phi\| = 1$. This proves that ϕ is completely bounded and that $\|\phi\|_{cb} = \|\phi\|$.

There is a corresponding result for bilinear maps which involves the Haagerup tensor product. Any bilinear map $\phi\colon \mathcal{E} \times \mathcal{E} \to B(H)$ on an operator subspace $\mathcal{E} \subseteq B(H)$ has an associated linear map $\tilde{\phi}\colon \mathcal{E} \otimes \mathcal{E} \to B(H)$ on the algebraic tensor product. If $\mathcal{E}$ is a bimodule with respect to a C^*-algebra $\mathcal{A}$ then we say that ϕ is $\mathcal{A}$-*multimodular* if

$$\phi(aeb, fc) = a\phi(e, bf)c, \qquad a, b, c \in \mathcal{A}, \quad e, f \in \mathcal{E},$$

with an obvious parallel definition for $\tilde{\phi}$:

$$\tilde{\phi}(aeb \otimes fc) = a\tilde{\phi}(e \otimes bf)c.$$

1.6.2 Theorem. *Let* $\phi\colon \mathcal{E} \times \mathcal{E} \to B(H)$ *be a bilinear map such that* $\tilde{\phi}$ *is bounded on* $\mathcal{E} \otimes^h \mathcal{E}$. *Moreover, suppose that* ϕ *is* $\mathcal{A}$-*multimodular for a* C^*-*algebra* $\mathcal{A}$ *satisfying one of the following hypotheses:*

(i) $\mathcal{A}$ *has a cyclic vector,*
(ii) *given* $\xi_1, \ldots, \xi_n \in H$ *there exists* $\xi \in H$ *such that* $\xi_i \in \overline{\mathcal{A}\xi}, 1 \leq i \leq n$.

Then ϕ *is completely bounded and* $\|\phi\|_{cb} = \|\tilde{\phi}\|$.

Proof. Assume that $\|\tilde{\phi}\| = 1$ and, to reach a contradiction, suppose that $\|\phi_n\| > 1$ for some $n \geq 1$. We may follow the proof of the previous theorem to obtain this situation: $(e_{ij}), (f_{ij}) \in \mathsf{M}_n(\mathcal{E})$ with norms at most 1, $\tilde{\xi}$, $\tilde{\eta} \in H$ with norms at most 1, $c_i, d_j \in \mathcal{A}$ such that $\sum_{i=1}^{n} c_i^* c_i, \sum_{j=1}^{n} d_j^* d_j \leq 1$, and

$$\left| \sum_{i,j,k} \langle \phi(e_{ik}, f_{kj}) d_j \tilde{\eta}, c_i \tilde{\xi} \rangle \right| > 1.$$

Using the multimodular properties of ϕ, this may be rewritten as

$$\left| \sum_{k} \left\langle \phi \left(\sum_{i} c_i^* e_{ik}, \sum_{j} f_{kj} d_j \right) \tilde{\eta}, \tilde{\xi} \right\rangle \right| > 1.$$

Let $e_k' = \sum_{i} c_i^* e_{ik}$, $f_k' = \sum_{j} f_{kj} d_j$. Then

$$(e_1', \ldots, e_n') = (c_1^*, \ldots, c_n^*)(e_{ij}), \qquad \begin{pmatrix} f_1' \\ \vdots \\ f_n' \end{pmatrix} = (f_{ij}) \begin{pmatrix} d_1 \\ \vdots \\ d_n \end{pmatrix}$$

from which it is easy to estimate that $\|(e_1', \ldots, e_n')\|$, $\left\| \begin{pmatrix} f_1' \\ \vdots \\ f_n' \end{pmatrix} \right\| \leq 1$. Thus,

setting $u = \sum_k e_k' \otimes f_k'$, we have $\|u\|_h \leq 1$ and

$$\|\tilde{\phi}(u)\| \geq |\langle \tilde{\phi}(u)\tilde{\eta}, \tilde{\xi} \rangle| > 1.$$

We have now contradicted $\|\tilde{\phi}\| = 1$ and so $\|\phi\|_{cb} \leq 1$.

Conversely, given $\varepsilon > 0$, we may find e_i, $f_i \in \mathcal{E}$ such that $\|(e_1, \ldots, e_n)\|$, $\left\| \begin{pmatrix} f_1 \\ \vdots \\ f_n \end{pmatrix} \right\| \leq 1$ and $\left\| \tilde{\phi} \left(\sum_i e_i \otimes f_i \right) \right\| \geq 1 - \varepsilon$. Let $A \in \mathsf{M}_n(\mathcal{E})$ have

$(e_1, \ldots, e_n)$ for its first row, $B \in \mathsf{M}_n(\mathcal{E})$ have $\begin{pmatrix} f_1 \\ \vdots \\ f_n \end{pmatrix}$ for its first column and

all other positions in both matrices are filled by 0's. Then $\|A\|, \|B\| \leq 1$ and

$$\|\phi_n(A, B)\| = \left\| \tilde{\phi} \left(\sum_i e_i \otimes f_i \right) \right\| \geq 1 - \varepsilon.$$

Thus $\|\phi\|_{cb} \geq 1 - \varepsilon$ for every $\varepsilon > 0$, proving the reverse inequality.

1.7 A Projection onto Completely Bounded Module Maps

Averaging is one of the fundamental computational techniques in simplifying cohomology calculations. In Chapter 3 averages over a compact or amenable group of unitaries are used to deduce the isomorphism of certain cohomology spaces (3.2.4). Both the compact and amenable cases are covered in Lemma 3.2.4. Projections can be considered as a type of average; conditional expectations in probability and von Neumann algebra theory are of this type. The projection constructed in the proof of Theorem 4.2.6 (see Lemma 4.2.1) when proving that the completely bounded cohomology into $B(H)$ is zero is another type of average. In this section a projection ρ is shown to exist from the space of completely bounded maps from a C^*-algebra $\mathcal{A}$ into a von Neumann algebra $\mathcal{M}$ onto the subspace of right $\mathcal{M}$-module maps (Theorem 1.7.4). Before the theorem is stated and the proof is given, some notation is introduced and several lemmas are proved.

If the von Neumann algebra $\mathcal{M}$ does not act on a separable Hilbert space, (equivalently does not have a separable predual), then it is essential to consider subsets indexed by larger sets than the integers below. To simplify things and to enable us to restrict attention to sequences, the proofs will be given only when $\mathcal{M}$ has a separable predual. The general case may be easily obtained by changing the index set from N to J.

Let $\mathcal{M}$ be a von Neumann algebra with separable predual. Let $\mathcal{P}$ be the set of all sequences $(m_j: j \in \mathbb{N})$ of elements from $\mathcal{M}$ such that $\sum m_j^* m_j$ converges strongly to 1 in $\mathcal{M}$. An advantage of this is that the m_j's can be chosen to be partial isometries v_j with $v_j v_j^* \leq e$ (a projection), $j \geq 2$, $\sum\limits_{j \geq 2} v_j^* v_j = c(e)$, the central support of e and $m_1 = 1 - c(e)$ (see the proof of Lemma 1.7.2). The following lemma contains the details of the construction of the semigroup $\mathbf{P}_\pi$ used in the proof. Sums are taken over $\mathbb{N}$ unless it is specified otherwise.

1.7.1 Lemma. *Let H be a separable Hilbert space, let $\mathcal{M}$ be a von Neumann subalgebra of a C^*-subalgebra $\mathcal{A}$ of $B(H)$ and let π be a representation of $\mathcal{A}$ on a Hilbert space K_π. Let $\mathbf{P}_\pi$ denote the set of all continuous linear maps α on $B(H, K_\pi)$ of the form $\alpha(T) = \sum \pi(m_j^*) T m_j$ with $m_j \in \mathcal{M}$ for all $j \in \mathbb{N}$ and $\sum m_j^* m_j = 1$ in the strong operator topology. Then $\mathbf{P}_\pi$ is a convex semigroup of normal maps on $B(H, K_\pi)$ with $\|\alpha\| \leq 1$ for all $\alpha \in \mathbf{P}_\pi$. Further $\alpha(T) = T$ for all $\alpha \in \mathbf{P}_\pi$ if and only if $\pi(m)T = Tm$ for all $m \in \mathcal{M}$.*

Proof. Recall that $\mathcal{P}$ denotes the set of all sequences (m_j) in $\mathcal{M}$ with $\sum m_j^* m_j = 1$ in the strong operator topology. The strong operator convergence of $\sum m_j^* m_j = 1$ corresponds to $\sum \|m_j \xi\|^2 = \|\xi\|^2$ for all $\xi \in H$ and implies that $\sum \|\pi(m_j)\eta\|^2 \leq \|\eta\|^2$ for all $\eta \in K_\pi$. For each finite set F in $\mathbb{N}$, $\xi \in H$ and $\eta \in K_\pi$,

$$(1) \quad \left| \left\langle \sum_F \pi(m_j^*) T m_j \xi, \eta \right\rangle \right|^2 \leq \left(\sum_F \|T\| \cdot \|m_j \xi\| \cdot \|\pi(m_j)\eta\| \right)^2$$

$$\leq \|T\|^2 \cdot \sum_F \|m_j \xi\|^2 \cdot \sum_F \|\pi(m_j)\eta\|^2.$$

This inequality applied to the tail of the sequence shows that $\sum \pi(m_j^*) T m_j$ converges in the weak operator topology to an operator denoted by $\alpha(T)$. The inequality (1) applied to $F = \{1, \ldots, n\}$ for all n shows that $\|\alpha(T)\| \leq \|T\|$ for all T in $B(H, K_\pi)$. Further, since the ultraweak and weak topologies coincide on the unit ball [T, p. 69] the series $\sum \pi(m_j^*) T m_j$ is ultraweakly convergent, since (1) implies that all the partial sums are bounded by 1 in norm.

The ultraweak–ultraweak continuity of α is obtained as follows. Let the ultraweak neighborhood of 0 in the image be

$$W = \left\{ W \in B(H, K_\pi): \left| \sum_i \langle W \xi_i^k, \eta_i^k \rangle \right| < \varepsilon \text{ for } 1 \leq k \leq n \right\}$$

determined by $\varepsilon > 0$, (ξ_i^k), (η_i^k), $1 \leq k \leq n$ with $\xi_i^k \in H$, $\eta_i^k \in K_\pi$ $(i \in \mathbf{N})$ and

$$\sum_i \|\xi_i^k\|^2 = 1 = \sum_i \|\eta_i^k\|^2 \quad \text{for } 1 \leq k \leq n.$$

Then

$$(2) \qquad \left| \sum_i \langle \alpha(T)\xi_i^k, \eta_i^k \rangle \right| = \left| \sum_{i,j} \langle \pi(m_j^*)Tm_j\xi_i^k\eta_i^k \rangle \right|$$

$$= \left| \sum_{i,j} \langle Tm_j\xi_i^k, \pi(m_j)\eta_i^k \rangle \right|.$$

Now

$$\sum_{i,j} \|m_j\xi_i^k\|^2 = \sum_{i,j} \langle m_j^*m_j\xi_i^k, \xi_i^k \rangle$$

$$= \sum_i \langle \xi_i^k, \xi_i^k \rangle$$

$$= 1,$$

and

$$\sum_{i,j} \|\pi(m_j)\eta_i^k\|^2 \leq 1$$

in a similar way. Reindexing the double sequences $(m_j\xi_i^k)$ and $(\pi(m_j)\eta_i^k)$ over i and j as a single sequence leads to elements in $\ell_2 \otimes H$ and $\ell_2 \otimes K_\pi$. Letting (2) be less than ε now defines the ultraweak neighborhood of 0 that α maps to the required neighbourhood $\mathcal{W}$. The normality of α is essentially a formality.

If (m_i) and (n_j) are in $\mathcal{P}$, then

$$\sum_{i,j} (n_j^*m_i)^*n_jm_i = \sum_i m_i^* \left(\sum_j n_j^*n_j \right) m_i = 1,$$

and

$$\sum_{i,j} \pi(n_jm_i)^*Tn_jm_i = \sum_i \pi(m_i^*) \left(\sum_j \pi(n_j^*)Tn_j \right) m_i,$$

with these series converging ultraweakly and the normality of α having been used in the last equality. Hence $\mathbf{P}_\pi$ is a convex semigroup.

The main equality obtained from the minimal compact convex $\mathbf{P}$-invariant set is in the following lemma.

1.7.2 Lemma. *Let $\mathcal{M}$ be a von Neumann algebra on a separable Hilbert space H and let $(\pi_\lambda: \lambda \in \Lambda)$ be a set of representations of $\mathcal{M}$ on Hilbert spaces K_λ. Let $K_\infty = \oplus_\lambda K_\lambda$, $\pi_\infty = \oplus_\lambda \pi_\lambda$ and let $H_\infty = \oplus_\lambda H_\lambda$ where H_λ is a copy of H corresponding to K_λ with the direct sum action of $\mathcal{M}$ on H_∞. Let $T_\lambda \in B(H_\lambda, K_\lambda)$ with $\|T_\lambda\| \leq 1$ for all λ, let $T = \oplus_\lambda T_\lambda \in B(H_\infty, K_\infty)$ and let $\mathcal{C}$ be the ultraweak closure of $\mathbf{P}_{\pi_\infty}(T_\infty)$ in $B(H_\infty, K_\infty)$. If $t = (t_\lambda)$ is in a minimal $\mathbf{P}_{\pi_\infty}$-invariant ultraweakly closed convex subset $\mathcal{C}_0$ of $\mathcal{C}$, then*

$$\|t_\lambda e\| = \|\pi_\lambda(e) t_\lambda e\| = \|t_\lambda c(e)\|$$

for all λ in Λ and all projections e in $\mathcal{M}$.

Proof. A standard maximality argument associated with the construction of the central support $c(e)$ of a projection e in $\mathcal{M}$ implies that there is a sequence (v_j) of partial isometries in $\mathcal{M}$ such that $v_k v_k^* \leq e$ for all k and $v_j^* v_j$ ($j \in \mathbf{N}$) are pairwise orthogonal projections with $\sum v_j^* v_j = c(e)$ [Di2, KR4, T]. Let (m_k) be the sequence $(1 - c(e), v_j: j \in \mathbf{N})$ in $\mathcal{P}$ with β_∞ the corresponding element of $\mathbf{P}_{\pi_\infty}$, so that

$$\beta_\infty(s) = \pi_\infty(1 - c(e)) s (1 - c(e)) + \sum \pi_\infty(v_j^*) s v_j$$

for all s in $B(H_\infty, K_\infty)$. Let β_λ be the corresponding element of $\mathbf{P}_{\pi_\lambda}$. Let z be a non-zero element in the centre $\mathcal{Z}$ of $\mathcal{M}$.

Suppose that for some λ_0 in Λ the inequality

(1) $$\|\beta_{\lambda_0}(t_{\lambda_0}) z\| < \|t_{\lambda_0} z\|$$

holds, where $t = (t_\lambda)$ is in $\mathcal{C}_0$. We shall show that

$$\mathcal{D} = \{(s_\lambda) \in \mathcal{C}_0 : \|s_{\lambda_0} z\| \leq \|\beta_{\lambda_0}(t_{\lambda_0}) z\|\}$$

is a non-empty ultraweakly closed convex subset of $\mathcal{C}_0$ invariant under the action of $\mathbf{P}_{\pi_\infty}$ and not equal to $\mathcal{C}_0$. The set $\mathcal{D}$ is non-empty, because $\beta_\infty(t)$ is in $\mathcal{D}$, and $\mathcal{D}$ is ultraweakly closed and convex, because of the nature of the inequality defining $\mathcal{D}$ in the λ_0 position. The $\mathbf{P}_{\pi_\infty}$-invariance of $\mathcal{D}$ follows from

$$\|\alpha(s)_{\lambda_0} z\| = \|\alpha_{\lambda_0}(s_{\lambda_0}) z\| = \|\alpha_{\lambda_0}(s_{\lambda_0} z)\| \leq \|s_{\lambda_0} z\| \leq \|\beta_{\lambda_0}(t_{\lambda_0}) z\|,$$

for $\alpha \in \mathbf{P}_{\pi_\infty}$. The second equality is valid because $z \in \mathcal{Z}$ commutes with the elements $n_j \in \mathcal{M}$ defining α. Note that $\mathcal{D}$ is not equal to $\mathcal{C}_0$, because t is in $\mathcal{C}_0$ but not in $\mathcal{D}$ by (1). These properties of $\mathcal{D}$ contradict the minimality of $\mathcal{C}_0$. Hence

(2) $$\|\beta_\lambda(t_\lambda) z\| = \|t_\lambda z\|$$

for all λ in Λ and z in $\mathcal{Z}$.

Apply equality (2) with $z = c(e)$ to yield

$$
(3) \qquad \|t_\lambda c(e)\| = \|\beta_\lambda(t_\lambda)c(e)\|
$$

$$
= \|\{\pi_\lambda(1 - c(e))t_\lambda(1 - c(e)) + \sum \pi_\lambda(v_j^*)t_\lambda v_j\}c(e)\|
$$

$$
= \|\sum \pi_\lambda(v_j^*)t_\lambda v_j\|
$$

$$
= \sup\{\|\pi_\lambda(v_j^*)t_\lambda v_j\|: j \in \mathbb{N}\}.
$$

The third equality is a consequence of $v_j c(e) = v_j$. The last equality holds because the supports and ranges of the operators $\pi_\lambda(v_j^*)t_\lambda v_j$ $(j \in \mathbb{N})$ are pairwise orthogonal since $v_j v_k^* = 0$ for $j \neq k$ follows from the orthogonality of the projections $v_j^* v_j$ and $v_k^* v_k$. This is just the observation that the norm of a diagonal matrix is the supremum of the norms of the diagonal entries.

The equalities $ev_j = v_j$ and $\|v_j\| = 1$ imply that

$$
\|\pi_\lambda(v_j^*)t_\lambda v_j\| \leq \|\pi_\lambda(e)t_\lambda e\|
$$

$$
\leq \|t_\lambda e\|
$$

$$
\leq \|t_\lambda c(e)\|
$$

for all j. This and (3) imply that

$$
\|t_\lambda c(e)\| \leq \|\pi_\lambda(e)t_\lambda e\| \leq \|t_\lambda(e)\| \leq \|t_\lambda c(e)\|
$$

proving the lemma.

In the proof of Theorem 1.7.4 we need to take a direct sum over "all" representations of a completely bounded operator from $\mathcal{A}$ into $\mathcal{M}$ up to suitable equivalence classes. Two representations $\phi(x) = U\pi(x)V$ and $\phi(x) = S\theta(x)T$ of a completely bounded linear operator ϕ from $\mathcal{A}$ into $B(H)$ are unitarily equivalent if there is a unitary operator W from K_π onto K_θ such that

$$
\theta(x) = W\pi(x)W^* \ (x \in \mathcal{A}), \ UW^* = S \text{ and } WV = T.
$$

Let $\mathbf{P}_\pi$ and $\mathbf{P}_\theta$ be the convex semigroups of operators defined in Lemma 1.7.1 associated with the equivalent representations π and θ above. Then

$$
\mathbf{P}_\theta(T) = \mathbf{P}_\theta(WV) = W\mathbf{P}_\pi(V)
$$

since $W\pi(x) = \theta(x)W \ (x \in \mathcal{A})$. This shows that it will be sufficient to work with equivalent representations of a completely bounded operator. The next lemma provides a bound on the dimension of the Hilbert space K_π.

1.7.3 Lemma. *Let H be a separable Hilbert space and let $\mathcal{A}$ be a C^*-algebra on H. If ϕ is a completely bounded linear operator from $\mathcal{A}$ into $B(H)$, then there is a representation π on a Hilbert space H_π of dimension no greater than c and continuous linear operators U_ϕ from K_π into H and V_ϕ from H into K_π such that $\phi(x) = U_\phi \pi(x) V_\phi$ for all x in $\mathcal{A}$ and $\|\phi\|_{cb} = \|U_\phi\| \cdot \|V_\phi\|$. If ϕ is normal, then π can be chosen to be normal.*

Proof. The representation theorem (1.3.1) for a completely bounded map ϕ from $\mathcal{A}$ into $B(H)$ gives a representation π_0 of $\mathcal{A}$ on a Hilbert space K_0 and continuous linear operators U_0 from K_0 into H and V_0 from H into K_0 such that $\phi(x) = U_0 \pi_0(A) V_0$ ($x \in \mathcal{A}$) and $\|\phi\|_{cb} = \|U_0\| \|V_0\|$. Let K_π be the closed linear span of the set $\pi_0(\mathcal{A}) V_0 H = \{\pi_0(a) V_0 \xi \colon a \in \mathcal{A}, \, \xi \in H\}$. Note that the dimension of K_π as a Hilbert space is no greater than

$$\dim(\mathcal{A}) \cdot \dim(H) \leq c \cdot \aleph_0 = c,$$

where $\dim(\mathcal{A})$ is the dimension of $\mathcal{A}$ as a Banach space, $\dim(H)$ is the dimension of H as a Hilbert space and c is the dimension of $B(H)$. Also note that K_π is a $\pi_0(\mathcal{A})$-invariant subspace of K_0. Let p be the projection from K_0 onto K_π, let π be the restriction of π_0 to K_π, let $V_\phi = pV_0$ and let U_ϕ be the restriction of U to K_π. These are the required representation and operators. Initially one obtains $\|\phi\|_{cb} \geq \|U_\phi\| \|V_\phi\|$ but the reverse inequality holds in general. If ϕ is normal, then π_0 can be chosen to be normal (1.5.7) so π is normal. This completes the proof.

1.7.4 Theorem. *Let H be a (separable) Hilbert space, let $\mathcal{A}$ be a C^*-subalgebra of $B(H)$ and let $\mathcal{M}$ be a von Neumann subalgebra of $B(H)$ contained in $\mathcal{A}$. There is a projection ρ from $\mathcal{L}_{cb}(\mathcal{A}, \mathcal{M})$ onto $\mathcal{L}_{cb}(\mathcal{A}, \mathcal{M})_\mathcal{M}$, the space of right $\mathcal{M}$-module maps in $\mathcal{L}_{cb}(\mathcal{A}, \mathcal{M})$, with $\|\rho\| \leq 1$. Further ρ has the following properties.:*

(1) The element $(\rho\phi)(x)$ is in the ultraweak closure of the convex set

$$\left\{ \sum \phi(xm_j^*)m_j \colon m_k \in \mathcal{M} \ (k \in \mathbb{N}), \sum m_j^* m_j = 1 \right\}$$

for all x in $\mathcal{A}$ and ϕ in $\mathcal{L}_{cb}(\mathcal{A}, \mathcal{M})$,

(2) For each representation π of $\mathcal{A}$ on a Hilbert space K_π of dimension no greater than c and each T in $B(H, K_\pi)$ with $\|T\| \leq 1$, there is a W in the ultraweak closure of

$$\left\{ \sum \pi(m_j^*)Tm_j \colon m_k \in \mathcal{M} \ (k \in \mathbb{N}), \sum m_j^* m_j = 1 \right\}$$

such that if $\phi \in \mathcal{L}_{cb}(\mathcal{A}, \mathcal{M})$ has a representation $\phi(x) = S\pi(x)T$ ($x \in \mathcal{A}$) for some S in $B(K_\pi, H)$ then $\rho\phi(x) = S\pi(x)W$ ($x \in \mathcal{A}$).

(3) The projection ρ maps $\mathcal{L}_{wcb}(\mathcal{A}, \mathcal{M})$ onto $\mathcal{L}_{wcb}(\mathcal{A}, \mathcal{M})_{\mathcal{M}}$.

Proof. The proof is given only when H is separable. Let Γ be a set of representatives for the set of all equivalence classes of equivalent representations of $\mathcal{A}$ on Hilbert spaces of cardinality no greater than c. Let Λ be the disjoint union

$$\cup\{\{T \in B(H, K_\pi): \|T\| = 1\}: \pi \in \Gamma\}.$$

In fact a much smaller set would suffice for the index set Λ. To each λ in Λ is associated a π in Γ and a T in $B(H, K_\pi)$ with $\|T\| = 1$; for notational convenience denote the associated Hilbert space, representation and operator by H_λ, π_λ and T_λ, respectively. Let $\mathcal{P}$ as usual denote the set of all sequences $(m_j: j \in \mathbf{N})$ from $\mathcal{M}$ with $\sum m_j^* m_j = 1$, let H_∞, K_∞ and π_∞ be defined as in Lemma 1.7.2 and let $\mathbf{P}_{\pi_\infty}$ be the semigroup of operators defined on $B(H_\infty, K_\infty)$ by $\alpha(S) = \sum \pi_\infty(n_j^*)Sn_j$ for $(n_j) \in \mathcal{P}$. Let $\mathcal{C}_0$ denote a minimal $\mathbf{P}_{\pi_\infty}$-invariant ultraweakly closed subset of the ultraweak closure $\mathcal{C}$ of $\mathbf{P}_{\pi_\infty}(T_\lambda)$ in $B(H_\infty, K_\infty)$. Note that a minimal $\mathcal{C}_0$ exists since $\mathcal{C}$ is non-empty and ultraweakly compact – it is an ultraweakly closed subset bounded in norm by 1 (see Lemma 1.7.1). By Lemma 1.7.2

$$(1) \qquad\qquad \|t_\lambda e\| = \|\pi_\lambda(e)t_\lambda e\| = \|t_\lambda c(e)\|$$

for all projections e in $\mathcal{M}$, all $t = (t_\lambda)$ in $\mathcal{C}_0$ and all λ in Λ. The aim is to prove that if ϕ in $\mathcal{L}_{cb}(\mathcal{A}, \mathcal{M})$ has a representation

$$\phi(x) = S_\lambda \pi_\lambda(x) T_\lambda \qquad (x \in \mathcal{A})$$

then $(\rho\phi)(x) = S_\lambda \pi_\lambda(x) t_\lambda$ $(x \in \mathcal{A})$ defines $\rho\phi$ in $\mathcal{L}_{cb}(\mathcal{A}, \mathcal{M})_{\mathcal{M}}$. The major part of the remainder of the proof is to show that $\rho\phi$ is in $\mathcal{L}_{cb}(\mathcal{A}, \mathcal{M})_{\mathcal{M}}$ by a contradiction argument.

Suppose there is a $\phi \in \mathcal{L}_{cb}(\mathcal{A}, \mathcal{M})$ with $\phi(x) = S_\lambda \pi_\lambda(x) T_\lambda$ $(x \in \mathcal{A})$ such that

$$(2) \qquad\qquad a = S_\lambda \pi_\lambda(x(1 - e))t_\lambda e \neq 0.$$

Then $a \in \mathcal{M}$ so a^*a is a non-zero element of $\mathcal{M}$. Let f be the spectral projection in $\mathcal{M}$ for a^*a corresponding to the closed interval $[\|a\|^2/2, \|a\|^2]$ and let

$$\varepsilon = \|a\|^2 (2\|S_\lambda \pi_\lambda(x)\| \cdot \|t_\lambda f\|)^{-2}.$$

The equation $ea^*ae = a^*a$ shows that $f \leq e$. Using equation (1) with the projection f, choose a unit vector η in fH_λ so that

$$(1 - \varepsilon)\|t_\lambda f\|^2 = (1 - \varepsilon)\|\pi_\lambda(f)t_\lambda f\|^2 \leq \|\pi_\lambda(f)t_\lambda f\eta\|^2.$$

Then

$$
\begin{aligned}
(1 - \varepsilon)\|t_\lambda f\|^2 &\leq \|\pi_\lambda(f)t_\lambda f\eta\|^2 \\
&\leq \|\pi_\lambda(e)t_\lambda f\eta\|^2 \\
&= \|t_\lambda f\eta\|^2 - \|(1 - \pi_\lambda(e))t_\lambda f\eta\|^2 \\
&\leq \|t_\lambda f\|^2 - \|(1 - \pi_\lambda(e))t_\lambda f\eta\|^2 \\
&\leq \|t_\lambda f\|^2 - \|S_\lambda \pi_\lambda(x)\|^{-2}\|S_\lambda \pi_\lambda(x(1 - e))t_\lambda f\eta\| \\
&= \|t_\lambda f\|^2 - \|S_\lambda \pi_\lambda(x)\|^{-2}\|a f\eta\|^2 \\
&\leq \|t_\lambda f\|^2 - \|S_\lambda \pi_\lambda(x)\|^{-2}\|a\|^2/2.
\end{aligned}
$$

(3)

The final inequality holds because f is the spectral projection of a^*a associated with the interval $[\|a\|^2/2, \|a\|^2]$ and $\eta = f\eta$ is a unit vector, showing that $\|a\|^2/2 \leq \|af\eta\|^2$. Now ε was chosen so that

$$
\|S_\lambda \pi_\lambda(x)\|^{-2} \cdot \|a\|^2/2 = 2\varepsilon\|t_\lambda f\|^2.
$$

Combining this inequality with (3) above yields

$$
(1 - \varepsilon)\|t_\lambda f\|^2 \leq (1 - 2\varepsilon)\|t_\lambda f\|^2.
$$

This is a contradiction since $t_\lambda f \neq 0$ and $\varepsilon > 0$; thus $a = 0$. Hence

(4) $$S_\lambda \pi_\lambda(x(1 - e))t_\lambda e = 0$$

for all projections $e \in \mathcal{M}$, all $x \in \mathcal{A}$, all $\phi \in \mathcal{L}_{cb}(\mathcal{A}, \mathcal{M})$ with $\phi(y) = S_\lambda \pi_\lambda(y)T_\lambda$ ($y \in \mathcal{A}$), and all $\lambda \in \Lambda$. Replacing e by $(1 - e)$ in (4) and subtracting the resulting equation from (4) gives

$$
S\pi(xe)t_\lambda = S\pi(x)t_\lambda e
$$

with the same quantifiers as in (4). From this it follows that

(5) $$S_\lambda \pi_\lambda(xm)t_\lambda = S_\lambda \pi_\lambda(x)t_\lambda m$$

for all $m \in \mathcal{M}$ because the projections span a von Neumann algebra.

It is now necessary to show that $\rho\phi$ is independent of the representation of ϕ chosen. Since $(t_\lambda) \in \mathcal{C}_0 \subseteq \mathcal{C}$, there is a net $(\alpha_i: i \in I)$ in $\mathbf{P}_{\pi_\infty}$ such that $\alpha_i(T_\lambda)$ tends ultraweakly to (t_λ). Let ϕ in $\mathcal{L}_{cb}(\mathcal{A}, \mathcal{M})$ have two representations as a completely bounded map

$$
\phi(x) = S_\pi \pi(x)T_\pi = S_\theta \theta(x)T_\theta \qquad (x \in \mathcal{A})
$$

with π and θ in Γ and $\|T_\pi\| = \|T_\theta\| = 1$. Let $\pi = \pi_\lambda$, $S_\pi = S_\lambda$, $T_\pi = T_\lambda$, and $\theta = \pi_\mu$, $S_\theta = S_\mu$ and $T_\theta = T_\mu$ where λ and μ are in Λ. If $\alpha \in \mathbf{P}_{\pi_\infty}$ corresponds to $\sum m_j^* m_j = 1$ with $m_k \in \mathcal{M}$ ($k \in \mathbb{N}$), then

$$(6) \qquad \sum \phi(x m_j^*) m_j = S_\lambda \pi_\lambda(x) \sum \pi_\lambda(m_j^*) T_\lambda m_j$$
$$= S_\mu \pi_\mu(x) \sum \pi_\mu(m_j^*) T_\mu m_j.$$

Replacing α by α_i and taking the ultraweak limit in (6) over the net $\mathcal{I}$ gives

$$(7) \qquad S_\lambda \pi_\lambda(x) t_\lambda = S_\mu \pi_\mu(x) t_\mu$$

for all $x \in \mathcal{A}$.

Define $\rho\phi(x) = S_\lambda \pi_\lambda(x) t_\lambda$ for all $x \in \mathcal{A}$ and all $\phi(y) = S_\lambda \pi_\lambda(y) T_\lambda$ ($y \in \mathcal{A}$) with λ in Λ. By (7), $\rho\phi(x)$ is well defined. By definition of $\mathcal{I}$, $\rho\phi(x)$ is the ultraweak limit of $\alpha_i(\phi)(x)$ for all $x \in \mathcal{A}$, where α_i corresponds to $\sum m_{j,i}^* m_{j,i} = 1$ in $\mathcal{M}$. Each of the maps α_i is in $\mathbf{P}_{\pi_\infty}$, so ρ is a linear operator from $\mathcal{L}_{cb}(\mathcal{A}, \mathcal{M})$ into itself with $\|\rho\| \leq 1$. Equation (5) implies that $\rho\phi(xm) = \rho\phi(x)m$ for all $x \in \mathcal{A}$, $m \in \mathcal{M}$ and $\phi \in \mathcal{L}_{cb}(\mathcal{A}, \mathcal{M})$; note that in representing ϕ we have to ensure $\|T\| \leq 1$. Properties (1) and (2) of the conclusion were observed above. If ϕ is in $\mathcal{L}_{cb}(\mathcal{A}, \mathcal{M})_\mathcal{M}$ and $m_k \in \mathcal{M}$ ($k \in \mathbb{N}$) with $\sum m_j^* m_j = 1$, then

$$\sum \phi(x m_j^*) m_j = \sum \phi(x) m_j^* m_j = \phi(x) \qquad (x \in \mathcal{A})$$

so $\rho\phi = \phi$. Hence ρ is a projection from $\mathcal{L}_{cb}(\mathcal{A}, \mathcal{M})$ onto $\mathcal{L}_{cb}(\mathcal{A}, \mathcal{M})_\mathcal{M}$.

Finally if $\phi \in \mathcal{L}_{wcb}(\mathcal{A}, \mathcal{M})$ then ϕ has a representation of the form $\phi(x) = S\pi(x)T$ ($x \in \mathcal{A}$) with π a normal representation in Γ by (1.5.7). By equation (7) and the definition of $\rho\phi$, $\rho\phi$ has a representation $(\rho\phi)(x) = S\pi(x)t_\pi$ ($x \in \mathcal{A}$) so is a normal operator. This proves Theorem 1.7.4.

Though the following result is not needed here, it has been used to show that completely complemented von Neumann algebras are injective. Recall that $\mathcal{L}_{cb}(\mathcal{A}, \mathcal{M}:/\mathcal{M})$ denotes the space of completely bounded $\mathcal{M}$-bimodule maps of $\mathcal{A}$ into $\mathcal{M}$.

1.7.5 Theorem. *Let $\mathcal{M}$ be a von Neumann subalgebra and let $\mathcal{A}$ be a C^*-subalgebra of $B(H)$ with $\mathcal{M}$ contained in $\mathcal{A}$. There is a projection β from the space $\mathcal{L}_{cb}(\mathcal{A}, \mathcal{M})$ of completely bounded operators from $\mathcal{A}$ into $\mathcal{M}$ onto the subspace $\mathcal{L}_{cb}(\mathcal{A}, \mathcal{M}:/\mathcal{M})$ of $\mathcal{M}$-bimodule maps. Further $\|\beta\| \leq 1$, β maps completely positive maps to completely positive maps, and β maps $\mathcal{L}_{wcb}(\mathcal{A}, \mathcal{M})$ onto $\mathcal{L}_{wcb}(\mathcal{A}, \mathcal{M}:/\mathcal{M})$. Also, for each $x \in \mathcal{A}$, and $\phi \in \mathcal{L}_{cb}(\mathcal{A}, \mathcal{M})$, $\beta\phi(x)$ is the ultraweak limit of elements of $\mathcal{M}$ of the form*

$$\sum_{i,j} m_i^* S\pi(m_i x n_j^*) T n_j$$

for (m_i) and (n_j) in $\mathcal{P}$.

Proof. Apply Theorem 1.7.4 to obtain $\rho\phi$, and define $\beta\phi = \rho((\rho\phi*)^*)$, where $\psi^*(x) = \psi(x^*)^*$ for all $x \in \mathcal{A}$. The operator $*$ on $\mathcal{L}_{cb}(\mathcal{A}, \mathcal{M})$ switches the left and right module actions. Hence $(\rho\phi^*)^*$ does not affect the left module structure so $\rho((\rho\phi^*)^*)$ is a two-sided $\mathcal{M}$-module map. If ϕ is completely positive with $\phi(x) = U^*\pi(x)U$ $(x \in \mathcal{A})$ and $\pi \in \Gamma$ and if $\rho\phi(x) = U^*\pi(x)W$ based on this representation, then $\phi = \phi^*$ and $\rho(\phi^*)^*(x) = W^*\pi(x)U$. Now using this representation gives $\rho(\rho(\phi^*)^*)(x) = W^*\pi(x)W$ because ϕ and $\rho(\phi^*)^*$ have the same representation and the same final bridging operator. The remaining property follows directly from the construction in the proof of Theorem 1.7.4.

1.8 Notes and Remarks

A general reference for this chapter is the book [Pa2] by Paulsen, which contains much more information on the theory of completely positive and completely bounded linear maps. The multilinear theory was surveyed in [ChS2]. Theorem 1.2.1 is due to Stinespring [Sti]. The other fundamental result of complete positivity is Theorem 1.2.3 which appeared in [Ar], although we give here a later proof from [SmW]. Matrix ordered spaces are defined in [CE1] which also contains the characterization of operator systems (Theorem 1.2.7).

Corollary 1.3.2 is Wittstock's Hahn–Banach theorem [W1] which started the theory of completely bounded maps. The analogue of Stinespring's theorem (Theorem 1.3.1) came later [Ha7, Pa2] although it is implicit in [W1]. The treatment we give here is due to Paulsen [Pa2].

The Haagerup tensor product was introduced by Haagerup, who called it the α-tensor product [Ha6]. It received its present name in [EK], and it has proved to be a powerful tool in the study of operator spaces [BP, BS, E3, Ha6, PS, Sm2]. We have included very little of the theory; just enough for the study of cohomology. Lemma 1.4.1 comes from [BP], Proposition 1.4.2 from [EK, Ha6] and Proposition 1.4.3 from [PS] although we present a later simpler proof from [BP]. The remaining results of this section may be found in [PS]. They are required for Theorem 1.5.4 which is the bilinear version of the Stinespring representation theorem. It was first proved for C^*-algebras [ChS1] and then generalized to operator spaces [PS]. We give the development from the latter paper.

The results on cyclic vectors in the sixth section are taken from [ChPSS] and [Sm2]. The basic ideas are, however, already present in [Ch5] and [Ha2]. The construction of the projection onto completely bounded module maps comes from [ChS5].

2

Derivations

2.1 Introduction

Let $\mathcal{A}$ be a C^*-algebra and let $\mathcal{V}$ be a Banach $\mathcal{A}$-bimodule, a Banach space with a bounded left and right action of $\mathcal{A}$. The two main examples of Banach $\mathcal{A}$-modules are $\mathcal{A}$ itself, and $B(H)$ when $\mathcal{A}$ is represented on H. A *derivation* δ: $\mathcal{A} \to \mathcal{V}$ is a linear map satisfying

$$\delta(ab) = a\delta(b) + \delta(a)b, \qquad a, b \in \mathcal{A},$$

or equivalently

$$a\delta(b) - \delta(ab) + \delta(a)b = 0.$$

A fixed element $v \in \mathcal{V}$ defines a derivation by

$$\delta(a) = av - va, \qquad a \in \mathcal{A},$$

and such a derivation is called *inner*.

The space of bounded derivations contains the subspace of inner derivations and the quotient vector space, regarded as an additive abelian group, is the first cohomology group of $\mathcal{A}$ with respect to the module $\mathcal{V}$, denoted $H^1(\mathcal{A}, \mathcal{V})$. The question of whether derivations are inner is then equivalent to determining whether the first cohomology group vanishes.

In this chapter we develop the basic theory of derivations on C^*-algebras. A derivation δ: $\mathcal{A} \to \mathcal{A}$ is bounded (Theorem 2.2.1) and is also ultraweakly continuous (Theorem 2.2.2). Every derivation then extends to a derivation on the ultraweak closure of $\mathcal{A}$, and this means that the study of derivations on C^*-algebras can largely be reduced to the von Neumann algebra case for suitable modules.

The third section concerns the extension of a derivation on a von Neumann algebra $\mathcal{M}$ to a larger type I von Neumann algebra, which is an important step in the theory. This is accomplished by the use of tensor products, and we include a brief discussion of this topic. The next section discusses derivations on hyperfinite von Neumann algebras with values in an arbitrary dual Banach module. Such algebras are ultraweak closures of increasing families of finite dimensional subalgebras and so we may average over compact unitary groups using Haar measure, a general technique in cohomology.

The main results of the chapter are contained in the fifth section. We prove the Kadison–Sakai theorem that every derivation of a von Neumann algebra to itself is inner, and then we present the result (due to Stampfli for $B(H)$ and to Zsido generally) that the minimum possible norm for the implementing operator can be achieved.

2.2 Continuity of Derivations

In this section we consider a derivation $\delta\colon \mathcal{A} \to \mathcal{A}$ where $\mathcal{A}$ is a C^*-algebra on a Hilbert space H. We will establish two fundamental results: δ is norm continuous and δ is ultraweakly continuous. We will then deduce that δ may be extended to an ultraweakly continuous derivation on the ultraweak closure $\bar{\mathcal{A}}$ of $\mathcal{A}$.

If a C^*-algebra $\mathcal{A}$ has a unit then

$$\delta(1) = \delta(1^2) = 1\delta(1) + \delta(1)1 = 2\delta(1)$$

and so $\delta(1) = 0$. Conversely, if $\mathcal{A}$ has no unit, then we may adjoin one and extend δ to the larger C^*-algebra by defining $\delta(1)$ to be 0. This allows us to restrict attention to unital C^*-algebras below. We say that δ is self-adjoint if

$$\delta(a^*) = \delta(a)^*, \qquad a \in \mathcal{A}.$$

The equation

$$2\delta(a) = [\delta(a) + \delta(a^*)^*] - i[i\delta(a) - i\delta(a^*)^*]$$

expresses any derivation as a linear combination of self-adjoint derivations.

2.2.1 Theorem. *Let $\mathcal{A}$ be a C^*-algebra and let $\delta\colon \mathcal{A} \to \mathcal{A}$ be a derivation. Then δ is bounded.*

Proof. From the preceding remarks we may assume that $\mathcal{A}$ is unital and that δ is self-adjoint. Then $\delta(a)$ is self-adjoint whenever a is self-adjoint.

We first consider a positive element a with positive square root b, and a state ϕ such that $\phi(a) = 0$. Then, using the Cauchy–Schwarz inequality,

$$
\begin{aligned}
|\phi(\delta(a))| &= |\phi(\delta(b^2))| \\
&= |\phi(b\delta(b)) + \phi(\delta(b)b)| \\
&\leq \phi(b^2)^{1/2}\phi(\delta(b)^2)^{1/2} + \phi(\delta(b)^2)^{1/2}\phi(b^2)^{1/2} \\
&= 0.
\end{aligned}
$$

Now let $x \in \mathcal{A}$ be a self-adjoint element and let ϕ be a state such that $|\phi(x)| = \|x\|$. Replacing x by $-x$ if necessary, we may assume that $\phi(x) = \|x\|$. Set

$$a = \|x\| - x \geq 0$$

and observe that $\phi(a) = 0$. The previous calculation then shows that

$$\phi(\delta(a)) = 0.$$

However, $\delta(a) = -\delta(x)$ since $\delta(1) = 0$, and so $\phi(\delta(x)) = 0$.

We will show that δ is bounded on the self-adjoint part $\mathcal{A}^h$ of $\mathcal{A}$ by using the closed graph theorem. To this end let $\{x_n\}_{n=1}^{\infty}$ be a sequence of self-adjoint elements converging in norm to 0, while $\{\delta(x_n)\}_{n=1}^{\infty}$ converges in norm to a self-adjoint element $y \in \mathcal{A}$. Choose states ϕ_n such that

$$|\phi_n(x_n + y)| = \|x_n + y\|, \qquad n \geq 1.$$

By the w^*-compactness of the state space, these states have an accumulation point ϕ and, by passing to a subsequence if necessary, we may assume that

$$\phi(y) = \lim_{n \to \infty} \phi_n(y), \quad \phi(\delta(y)) = \lim_{n \to \infty} \phi_n(\delta(y)).$$

Then

$$|\phi(y)| = \lim_{n \to \infty} |\phi_n(x_n + y)| = \lim_{n \to \infty} \|x_n + y\| = \|y\|$$

and so $\phi(\delta(y)) = 0$ from above. Also

$$\phi_n(\delta(x_n) + \delta(y)) = 0$$

since $|\phi_n(x_n + y)| = \|x_n + y\|$. Thus

$$\phi(y) = \phi(y + \delta(y)) = (\phi - \phi_n)(y + \delta(y)) + \phi_n(y + \delta(y))$$
$$= (\phi - \phi_n)(y + \delta(y)) + \phi_n(\delta(x_n) + \delta(y)) + \phi_n(y - \delta(x_n))$$

for $n \geq 1$. Let $n \to \infty$ to obtain $\phi(y) = 0$. Thus $y = 0$, and so δ is continuous on $\mathcal{A}^h$ by the closed graph theorem. Since $\mathcal{A} = \mathcal{A}^h + i\mathcal{A}^h$, it follows that δ is bounded on $\mathcal{A}$.

2.2.2 Theorem. *Let $\mathcal{A}$ be a C^*-algebra on a Hilbert space H and let $\delta \colon \mathcal{A} \to \mathcal{A}$ be a derivation. Then δ is ultraweakly continuous, and extends to a derivation on the ultraweak closure $\bar{\mathcal{A}}$ of $\mathcal{A}$.*

Proof. Let $\mathcal{A}_1, \mathcal{A}_1^h$ and $\mathcal{A}_1^+$ denote respectively the closed unit ball of $\mathcal{A}$, the self-adjoint part of $\mathcal{A}_1$, and the positive part of $\mathcal{A}_1$. Fix two vectors $\xi, \eta \in H$ and consider $a \geq 0$ with positive square root b. Then

$$|\langle \delta(a)\xi, \eta \rangle| = |\langle \delta(b^2)\xi, \eta \rangle|$$
$$= |\langle (b\delta(b) + \delta(b)b)\xi, \eta \rangle|$$
$$\leq |\langle \delta(b)\xi, b\eta \rangle| + |\langle \delta(b)b\xi, \eta \rangle|$$
$$\leq \|\delta\| \, \|b\| \, \|\xi\| \, \|b\eta\| + \|\delta\| \, \|b\| \, \|b\xi\| \, \|\eta\|$$

since δ is bounded by (2.2.1). Now $\|b\|^2 = \|a\|$, and

$$\|b\xi\|^2 = \langle b\xi, b\xi \rangle = \langle a\xi, \xi \rangle \leq \|a\xi\| \|\xi\|$$

with a similar estimate for ξ replaced by η. Thus

$$|\langle \delta(a)\xi, \eta \rangle| \leq \|\delta\| \|a\|^{1/2}(\|\xi\| \|\eta\|^{1/2}\|a\eta\|^{1/2} + \|\eta\| \|\xi\|^{1/2}\|a\xi\|^{1/2}),$$

showing that the map

$$a \to \langle \delta(a)\xi, \eta \rangle$$

is continuous at 0 on $\mathcal{A}_1^+$ in the relative strong topology.

Each self-adjoint element $a \in \mathcal{A}_1^h$ may be written $a = a^+ - a^-$ where a^+, $a^- \in \mathcal{A}_1^+$ and $a^+a^- = 0$. Then $a^+\xi$ and $a^-\xi$ are orthogonal vectors for each $\xi \in H$ and so

$$\|a^+\xi\|, \|a^-\xi\| \leq \|a^+\xi - a^-\xi\| = \|a\xi\|.$$

Thus the maps $a \to a^+$, $a \to a^-$, are strongly continuous at 0 on $\mathcal{A}_1^h$, and so

$$a \to \langle \delta(a)\xi, \eta \rangle = \langle \delta(a^+)\xi, \eta \rangle - \langle \delta(a^-)\xi, \eta \rangle$$

is strongly continuous at 0 on $\mathcal{A}_1^h$. A net $\{a_\lambda\}$ from $\mathcal{A}_1^h$ converges strongly to $a \in \mathcal{A}_1^h$ if and only if $\{(a_\lambda - a)/2\}$ is a net in $\mathcal{A}_1^h$ converging strongly to 0. It follows that the map $a \to \langle \delta(a)\xi, \eta \rangle$ is strongly continuous on $\mathcal{A}_1^h$. Thus the inverse image of a closed convex subset of $\mathbb{C}$ is a convex relatively strongly closed subset of $\mathcal{A}_1^h$, which is also closed in the relative weak operator topology since convex subsets of $B(H)$ have identical strong and weak operator closures. We conclude that $a \to \langle \delta(a)\xi, \eta \rangle$ is continuous on $\mathcal{A}_1^h$ in the relative weak operator topology because the topology of closed sets in $\mathbb{C}$ is generated by the closed convex sets.

The adjoint map $a \to a^*$ is weak operator continuous on $\mathcal{A}$, so we see that

$$a \to \langle \delta(a)\xi, \eta \rangle = \left\langle \delta\left(\frac{a + a^*}{2}\right)\xi, \eta \right\rangle + i\left\langle \delta\left(\frac{a - a^*}{2i}\right)\xi, \eta \right\rangle$$

is continuous on $\mathcal{A}_1$ in the relative weak operator topology. The ultraweak and the weak operator topology coincide on $\mathcal{A}_1$ and so the restriction of δ to $\mathcal{A}_1$ is ultraweakly continuous. A simple limit argument then allows us to extend δ to an ultraweakly continuous map $\bar{\delta}$ on the ultraweak closure $\bar{\mathcal{A}}$ of $\mathcal{A}$. Multiplication in $\bar{\mathcal{A}}$ is separately ultraweakly continuous in each variable, from which it follows that $\bar{\delta}$ is also a derivation.

This result will enable us to restrict attention to von Neumann algebras, rather than general C^*-algebras. The next few technical lemmas will aid our investigation of derivations on von Neumann algebras.

2.2.3 Lemma. *Let* $\delta: \mathcal{A} \to \mathcal{A}$ *be a derivation on a C^*-algebra and let $\mathcal{J}$ be a closed two-sided ideal in $\mathcal{A}$. Then δ maps $\mathcal{J}$ into itself.*

Proof. Let a be a positive element of $\mathcal{J}$ with positive square root $b \in \mathcal{J}$. Then

$$\delta(a) = \delta(b^2) = b\delta(b) + \delta(b)b \in \mathcal{J}.$$

Since $\mathcal{J}$ is the span of its positive elements, the result follows.

2.2.4 Lemma. *Let* $\delta: \mathcal{A} \to \mathcal{A}$ *be a derivation on a C^*-algebra $\mathcal{A}$. Then δ annihilates the centre $\mathcal{Z}$.*

Proof. By (2.2.2) we may extend δ to a derivation on the ultraweak closure $\bar{\mathcal{A}}$ of $\mathcal{A}$ in some faithful representation of $\mathcal{A}$, and $\mathcal{Z}$ is contained in the centre of $\bar{\mathcal{A}}$. This allows us to restrict to the case when $\mathcal{A}$ is a von Neumann algebra and so is the norm closed span of its projections. It suffices to show that $\delta(p) = 0$ for each central projection p.

By (2.2.3) δ maps the ideal $\mathcal{A}p$ to itself, and p is the identity element of this unital algebra. From the calculation at the beginning of this section, we see that $\delta(p) = 0$. Thus δ annihilates $\mathcal{Z}$.

For $a \in \mathcal{A}$ and $z \in \mathcal{Z}$, the derivation equations

$$\delta(az) = \delta(a)z + a\delta(z),$$
$$\delta(za) = \delta(z)a + z\delta(a)$$

reduce to

$$\delta(az) = \delta(a)z, \quad \delta(za) = z\delta(a)$$

from (2.2.4). We will refer to any linear map which obeys these equations as a $\mathcal{Z}$-bimodule map. We investigate such maps in the next section.

2.3 Extension of Derivations

Let $\mathcal{M} \subseteq B(H)$ be a von Neumann algebra with centre $\mathcal{Z}$ and commutant $\mathcal{M}'$. We fix an arbitrary maximal abelian subalgebra (masa) $\mathcal{B}$ of $\mathcal{M}'$, which clearly contains $\mathcal{Z}$. For subsequent developments it is important to extend a derivation $\delta: \mathcal{M} \to \mathcal{M}$ to a derivation on the larger C^*-algebra $C^*(\mathcal{M}, \mathcal{B})$ generated by $\mathcal{M}$ and $\mathcal{B}$. To accomplish this we begin with a brief review of the minimal tensor product (see Section 5.2 for an alternative treatment of this extension). The details may be found in [T], or in any of the standard texts.

Consider two C^*-algebras $\mathcal{A}_1 \subseteq B(H)$ and $\mathcal{A}_2 \subseteq B(K)$. Then $\mathcal{A}_1 \otimes \mathcal{A}_2$ denotes the algebra generated by the operators $\{a_1 \otimes a_2 \in B(H \otimes K): a_1 \in \mathcal{A}_1, a_2 \in \mathcal{A}_2\}$. The norm closure of $\mathcal{A}_1 \otimes \mathcal{A}_2$ is denoted by $\mathcal{A}_1 \otimes_{\min} \mathcal{A}_2$

and is a C^*-algebra called the *minimal tensor product*. If both algebras are von Neumann algebras then $\mathcal{A}_1 \overline{\otimes} \mathcal{A}_2$ denotes the ultraweak closure of $\mathcal{A}_1 \otimes \mathcal{A}_2$, called the *spatial von Neumann algebra tensor product*. Both constructions appear to depend on the particular spatial representations of $\mathcal{A}_1$ and $\mathcal{A}_2$, but this is not the case. The norm $\|\cdot\|_{\min}$ on $\mathcal{A}_1 \otimes_{\min} \mathcal{A}_2$ is the minimal cross norm for which $\mathcal{A}_1 \otimes \mathcal{A}_2$ is a pre-C^*-algebra. The *maximal C^*-norm* $\|\cdot\|_{\max}$ on $\mathcal{A}_1 \otimes \mathcal{A}_2$ is also important.

We consider all $*$-homomorphisms $\pi\colon \mathcal{A}_1 \otimes \mathcal{A}_2 \to B(H)$ (with H varying) and define

$$\|u\|_{\max} = \sup_\pi \|\pi(u)\|, \qquad u \in \mathcal{A}_1 \otimes \mathcal{A}_2.$$

The maximal C^*-tensor product $\mathcal{A}_1 \otimes_{\max} \mathcal{A}_2$ is the completion of $\mathcal{A}_1 \otimes \mathcal{A}_2$ in this norm. As the names imply, any C^*-norm $\|\cdot\|_\alpha$ on $\mathcal{A}_1 \otimes \mathcal{A}_2$ satisfies

$$\|\cdot\|_{\min} \leq \|\cdot\|_\alpha \leq \|\cdot\|_{\max}.$$

In general the maximal and minimal norms are distinct, but they coincide in the important special case when one of the algebras is abelian. Then there is a unique C^*-norm on $\mathcal{A}_1 \otimes \mathcal{A}_2$. If $\mathcal{A}_2$ is isomorphic to the algebra $C(\Omega)$ of continuous functions on a compact Hausdorff space Ω then there is a natural map π of $\mathcal{A}_1 \otimes C(\Omega)$ into the C^*-algebra $C(\Omega, \mathcal{A}_1)$ of $\mathcal{A}_1$-valued continuous functions on Ω:

$$\pi(a \otimes f)(\omega) = f(\omega)a, \qquad a \in \mathcal{A}, \quad \omega \in \Omega.$$

This induces a C^*-norm on $\mathcal{A}_1 \otimes C(\Omega)$ which must be $\|\cdot\|_{\min}$, and so

$$\pi\colon \mathcal{A}_1 \otimes_{\min} C(\Omega) \to C(\Omega, \mathcal{A}_1)$$

is an isometric $*$-isomorphism. It is not difficult to see that it is also surjective. When convenient we will identify $\mathcal{A}_1 \otimes_{\min} C(\Omega)$ with $C(\Omega, \mathcal{A}_1)$. We now return to derivations.

2.3.1 Lemma. *Let $\delta\colon \mathcal{M} \to \mathcal{M}$ be a derivation and let $\mathcal{B}$ be an abelian C^*-algebra. Then there is a bounded derivation $\delta \otimes I\colon \mathcal{M} \otimes_{\min} \mathcal{B} \to \mathcal{M} \otimes_{\min} \mathcal{B}$ defined on elementary tensors by*

$$(\delta \otimes I)(m \otimes b) = \delta(m) \otimes b, \qquad m \in \mathcal{M}, \quad b \in \mathcal{B},$$

and $\|\delta \otimes I\| = \|\delta\|$.

Proof. Identify $\mathcal{B}$ with $C(\Omega)$ for some compact Hausdorff space Ω, and define $\delta_1\colon C(\Omega, \mathcal{M}) \to C(\Omega, \mathcal{M})$ by

$$\delta_1(f)(\omega) = \delta(f(\omega)), \qquad \omega \in \Omega, \quad f \in C(\Omega, \mathcal{M}).$$

It is easy to check that δ_1 is a derivation, and that $\|\delta_1\| = \|\delta\|$. Under the identification of $\mathcal{M} \otimes_{\min} C(\Omega)$ with $C(\Omega, \mathcal{M})$, δ_1 corresponds to $\delta \otimes I$ since, for $m \in \mathcal{M}$ and $g \in C(\Omega)$,

$$\pi((\delta \otimes I)(m \otimes g))(\omega) = \pi(\delta(m) \otimes g)(\omega)$$
$$= g(\omega)\delta(m)$$

while
$$\delta_1(\pi(m \otimes g))(\omega) = \delta(g(\omega)m)$$
$$= g(\omega)\delta(m).$$

Thus $\delta \otimes I$ is a bounded derivation and $\|\delta \otimes I\| = \|\delta_1\| = \|\delta\|$.

2.3.2 Theorem. *Let* $\delta\colon \mathcal{M} \to \mathcal{M}$ *be a derivation, and let* $\mathcal{B}$ *be a maximal abelian subalgebra of* $\mathcal{M}'$. *Then there exists a bounded derivation* $\tilde{\delta}\colon C^*(\mathcal{M}, \mathcal{B}) \to C^*(\mathcal{M}, \mathcal{B})$ *which is a norm preserving extension of* δ.

Proof. Since $\mathcal{M}$ and $\mathcal{B}$ are commuting algebras, the multiplication map $\mu\colon \mathcal{M} \otimes \mathcal{B} \to B(H)$ defined by

$$\mu\left(\sum_{i=1}^n m_i \otimes b_i\right) = \sum_{i=1}^n m_i b_i$$

is a $*$-homomorphism on $\mathcal{M} \otimes \mathcal{B}$, and is thus bounded on $\mathcal{M} \otimes_{\max} \mathcal{B}$ by definition of the maximal norm. In this case the maximal and minimal norms agree because $\mathcal{B}$ is abelian and so μ is bounded on $\mathcal{M} \otimes_{\min} \mathcal{B}$. The range of μ is norm dense in $C^*(\mathcal{M}, \mathcal{B})$ and is thus equal to this C^*-algebra. We then have an induced $*$-isomorphism $\tilde{\mu}\colon (\mathcal{M} \otimes_{\min} \mathcal{B})/\ker \mu \to C^*(\mathcal{M}, \mathcal{B})$. The induced derivation $\delta \otimes I$ on $\mathcal{M} \otimes_{\min} \mathcal{B}$ of (2.3.1) leaves invariant the ideal $\ker \mu$, by (2.2.3), and so allows us to define δ_1 on $(\mathcal{M} \otimes_{\min} \mathcal{B})/\ker \mu$ by

$$\delta_1(m \otimes b + \ker \mu) = \delta(m) \otimes b + \ker \mu.$$

Clearly $\|\delta_1\| \leq \|\delta \otimes I\| = \|\delta\|$. Define $\tilde{\delta}$ on $C^*(\mathcal{M}, \mathcal{B})$ by $\tilde{\delta} = \tilde{\mu}\delta_1\tilde{\mu}^{-1}$. Then $\tilde{\delta}$ is a derivation and

$$\tilde{\delta}(mb) = \tilde{\mu}\delta_1(m \otimes b + \ker \mu)$$
$$= \tilde{\mu}(\delta(m) \otimes b + \ker \mu)$$
$$= \delta(m)b.$$

Then $\tilde{\delta}$ extends δ (by taking $b = 1$) and so $\|\tilde{\delta}\| \geq \|\delta\|$ which, combined with $\|\delta_1\| \leq \|\delta\|$, shows that $\tilde{\delta}$ is a norm preserving extension.

2.4 Hyperfinite von Neumann Algebras

We say that a von Neumann algebra $\mathcal{M}$ is *hyperfinite* if there is an increasing net of unital finite dimensional subalgebras $\{\mathcal{M}_\lambda\}_{\lambda \in \Lambda}$ such that $\bigcup_\lambda \mathcal{M}_\lambda$ is ultraweakly dense in $\mathcal{M}$. Each derivation $\delta \colon \mathcal{M} \to \mathcal{M}$ induces a family of derivations $\delta_\lambda \colon \mathcal{M}_\lambda \to \mathcal{M}$ by restriction, and so we begin by considering derivations on finite dimensional subalgebras with values in $\mathcal{M}$.

2.4.1 Lemma. *Let $\mathcal{F}$ be a finite dimensional C^*-algebra. Then there exists a finite group G of unitary elements whose span is $\mathcal{F}$.*

Proof. Finite dimensional C^*-algebras are finite direct sums of matrix algebras and so it suffices to consider the case of a matrix algebra $\mathbf{M}_n$, with matrices written relative to the standard basis $\{\xi_i\}_{i=1}^n$ for $\mathbf{C}^n$. Let G_1 be the finite group of diagonal matrices whose entries are ± 1, and let G_2 be the finite group of unitary matrices which permute the basis. If $u \in G_1$ and $v \in G_2$ then $vuv^* \in G_1$ so $vu = wv$ for some $w \in G_1$. Thus $G = \{uv \colon u \in G_1, v \in G_2\}$ is a finite group of unitary matrices, and the linear span of G is a self-adjoint unital subalgebra $\mathcal{A}$ of $\mathbf{M}_n$. The group G_1 spans the algebra of diagonal matrices and so $\mathcal{A}$ contains the matrix unit E_{11}. For $1 \le k \le n$ let $u_k \in G_2$ be the permutation matrix which interchanges ξ_1 and ξ_k while leaving the remaining basis vectors fixed. Then $u_i E_{11} u_j$ is the matrix unit E_{ij} and so $\mathcal{A} = \mathbf{M}_n$.

2.4.2 Proposition. *Let $\mathcal{F}$ be a finite dimensional C^*-algebra, let $\mathcal{V}$ be a Banach $\mathcal{F}$-bimodule, and let $\delta \colon \mathcal{F} \to \mathcal{V}$ be a bounded derivation. Then there exists an element $v \in \mathcal{V}$ such that*

$$\delta(a) = av - va \quad \text{for} \quad a \in \mathcal{F},$$

and $\|v\| \le \|\delta\|$.

Proof. From (2.4.1) there is a finite group G of unitaries in $\mathcal{F}$ which spans $\mathcal{F}$. Let $u_1, \ldots, u_k$ be a listing of the elements of G, and define

$$v = k^{-1} \sum_{i=1}^k u_i^* \delta(u_i) \in \mathcal{V}.$$

Then $\|v\| \le \|\delta\|$. Now, for $1 \le i, j \le k$,

$$\delta(u_i u_j) = u_i \delta(u_j) + \delta(u_i) u_j$$

so

$$\delta(u_j) + u_i^* \delta(u_i) u_j = u_i^* \delta(u_i u_j) = u_j (u_i u_j)^* \delta(u_i u_j).$$

Divide by k and sum over i while keeping j fixed to obtain

$$\delta(u_j) + vu_j = u_j v$$

since $\{u_i u_j\}_{i=1}^k$ is another listing of the elements of G. Thus

$$\delta(u) = uv - vu, \qquad u \in G.$$

However G spans $\mathcal{F}$, so

$$\delta(a) = av - va, \qquad a \in \mathcal{F}.$$

We are now in a position to address the problem of whether derivations on von Neumann algebras are inner, at least in the hyperfinite case. We prove something stronger.

2.4.3 Theorem. *(i) Let $\mathcal{M} \subseteq \mathcal{N}$ be von Neumann algebras and let $\delta \colon \mathcal{M} \to \mathcal{N}$ be a bounded derivation. If $\mathcal{M}$ is hyperfinite, then there exists an element $n \in \mathcal{N}$ such that*

$$\delta(x) = xn - nx, \qquad x \in \mathcal{M}$$

and $\|n\| \le \|\delta\|$.

(ii) If $\mathcal{M}$ is hyperfinite and $\mathcal{V}$ is a dual normal $\mathcal{M}$-bimodule then a bounded derivation $\delta \colon \mathcal{M} \to \mathcal{V}$ is inner.

Proof. (i) Let $\{\mathcal{F}_\lambda\}_{\lambda \in \Lambda}$ be an increasing net of finite dimensional subalgebras of $\mathcal{M}$ whose union is ultraweakly dense in $\mathcal{M}$, and let $\delta_\lambda \colon \mathcal{F}_\lambda \to \mathcal{N}$ be the restriction of δ to $\mathcal{F}_\lambda$. By (2.4.2) there exist elements $n_\lambda \in \mathcal{N}$, $\|n_\lambda\| \le \|\delta\|$, such that

$$\delta_\lambda(x) = xn_\lambda - n_\lambda x, \qquad x \in \mathcal{F}_\lambda.$$

The ball in $\mathcal{N}$ of radius $\|\delta\|$ is ultraweakly compact and so $\{n_\lambda\}_{\lambda \in \Lambda}$ has a co-final ultraweakly convergent subnet $\{n_\omega\}_{\omega \in \Omega}$ with limit $n \in \mathcal{N}$. If $x \in \bigcup_\lambda \mathcal{F}_\lambda$ then a simple limit argument shows that

$$\delta(x) = xn - nx.$$

Thus δ is implemented by n on an ultraweakly dense subalgebra. The ultraweak continuity of δ (2.2.2) then gives the same conclusion on $\mathcal{M}$.

(ii) In (i) we only used the property of $\mathcal{N}$ that its norm closed balls are w^*-compact. Thus the same proof applies to this more general case.

We single out one important consequence, for the case when $\mathcal{M}$ itself is not hyperfinite but contains a hyperfinite subalgebra.

2.4.4 Corollary. *Let $A \subseteq M \subseteq N$ be von Neumann algebras, where A is hyperfinite. If $\delta\colon M \to N$ is a derivation then there exists an inner derivation δ_1 such that $\delta - \delta_1$ annihilates A.*

Proof. By (2.4.3) the restriction of δ to A is implemented by an element $n \in N$. Define $\delta_1\colon M \to N$ by

$$\delta_1(x) = xn - nx, \qquad x \in M.$$

Then $\delta - \delta_1$ annihilates A.

Direct sums and tensor products of hyperfinite von Neumann algebras are again hyperfinite. In particular, direct sums of algebras $Z \overline{\otimes} B(H)$, where Z is an abelian von Neumann algebra, are hyperfinite. Such direct sums constitute the class of type I von Neumann algebras and so the following important result is a special case of (2.4.3).

2.4.5 Corollary. *If $M \subseteq N$ are von Neumann algebras and M is type I, then every derivation $\delta\colon M \to N$ is implemented by an element n of N. This element may be chosen to satisfy $\|n\| \le \|\delta\|$.*

We conclude this section by showing that derivations on von Neumann algebras are spatially implemented.

2.4.6 Theorem. *Let $M \subseteq B(H)$ be a von Neumann algebra and let $\delta\colon M \to M$ be a derivation. Then there exists an operator $t \in B(H)$ such that $\|t\| \le \|\delta\|$ and*

$$\delta(x) = xt - tx, \qquad x \in M.$$

Proof. Fix an arbitrary masa $B \subseteq M'$ and extend δ to a derivation $\tilde{\delta}\colon C^*(M, B) \to C^*(M, B)$ by (2.3.2). Theorem 2.2.2 then allows us to extend $\tilde{\delta}$ to a derivation δ_1 on the ultraweak closure $W^*(M, B)$ of $C^*(M, B)$. Both extensions are norm preserving and so $\|\delta_1\| = \|\delta\|$.

Any operator $x \in B(H)$ which commutes with $W^*(M, B)$ must commute with M, so lies in M'. Since B is maximal abelian in M', it follows that $x \in B$. On the other hand, it is clear that any element of B commutes with $W^*(M, B)$, which establishes that $W^*(M, B)' = B$. A von Neumann algebra whose commutant is abelian must be type I [KR4]. We may now apply (2.4.5) to $W^*(M, B)$ to conclude that δ_1 is inner. Thus δ, the restriction of δ_1 to M, is spatially implemented by an operator $t \in W^*(M, B)$ with $\|t\| \le \|\delta\|$.

2.5 Inner Derivations

In Theorem 2.4.6 we showed that a derivation $\delta\colon \mathcal{M} \to \mathcal{M}$ on a von Neumann algebra $\mathcal{M} \subseteq B(H)$ is implemented by an operator $t \in B(H)$ with $\|t\| \le \|\delta\|$:

$$\delta(m) = mt - tm, \qquad m \in \mathcal{M}.$$

We can now improve this result by showing that t can be chosen to lie in $\mathcal{M}$, so that δ is an inner derivation.

2.5.1 Theorem. Let $\delta\colon \mathcal{M} \to \mathcal{M}$ be a derivation on a von Neumann algebra $\mathcal{M} \subseteq B(H)$. Then there exists $m_0 \in \mathcal{M}$, $\|m_0\| \le \|\delta\|$, such that

$$\delta(m) = mm_0 - m_0 m, \qquad m \in \mathcal{M}.$$

Proof. Choose $t \in B(H)$, $\|t\| \le \delta$ such that

$$\delta(m) = mt - tm, \qquad m \in \mathcal{M}.$$

This equation may be rewritten

$$\delta(m) = (1\ t) \begin{pmatrix} m & 0 \\ 0 & m \end{pmatrix} \begin{pmatrix} t \\ -1 \end{pmatrix}, \qquad m \in \mathcal{M},$$

and so has the form $S\pi(m)T$ where π is the 2-fold amplification of the identity representation. By Theorem 1.7.4, the image $\rho\delta$ of δ under the projection ρ onto right $\mathcal{M}$-module maps is expressible by

$$\rho\delta(m) = S\pi(m)W, \qquad m \in \mathcal{M},$$

with W in the ultraweakly closed convex hull of $\{\sum \pi(m_i)Tm_i^*\colon m_i \in \mathcal{M}, \sum m_i m_i^* = 1\}$. Since $T = \begin{pmatrix} t \\ -1 \end{pmatrix}$, W must have the form $\begin{pmatrix} b \\ -1 \end{pmatrix}$. Then

$$\rho\delta(m) = mb - tm, \qquad m \in \mathcal{M}.$$

Moreover, the map $m \to \rho\delta(m) + tm$ is a right $\mathcal{M}$-module map which forces b to lie in $\mathcal{M}'$. Putting $m = 1$, we find that

$$b - t = \rho\delta(1) \in \mathcal{M}$$

so that t has the form $b + m_0$ for some $m_0 \in \mathcal{M}$. Since ρ is a contraction,

$$\|m_0\| = \| - \rho\delta(1)\| \le \|\delta\|,$$

and m_0 implements δ because we have already shown that $b \in \mathcal{M}'$.

It is clear that an element m_0 which implements an inner derivation δ must have norm at least $\frac{1}{2}\|\delta\|$, so our last result shows that every derivation δ on a von Neumann algebra $\mathcal{M}$ is implemented by an element $m_0 \in \mathcal{M}$ satisfying

$$\frac{1}{2}\|\delta\| \leq \|m_0\| \leq \|\delta\|.$$

The element m_0 is not unique because we may perturb by any element of the centre $\mathcal{Z}(\mathcal{M})$. We close this section by showing that a suitable central perturbation of m_0 allows us to achieve the lower bound $\frac{1}{2}\|\delta\|$. We will examine successively finite matrix algebras, $B(H)$, and then the general case.

2.5.2 Lemma. *Let $T \in \mathsf{M}_n$ and let δ be the inner derivation implemented by T. Then*

$$\|\delta\| = 2 \inf_{\lambda \in \mathbb{C}} \|T - \lambda T\| = 2 \operatorname{dist}(T, \mathbb{C}I).$$

Proof. By first subtracting the closest element of $\mathbb{C}I$ to T, we may assume that $\operatorname{dist}(T, \mathbb{C}I) = \|T\|$. Then we must prove that $\|\delta\| = 2\|T\|$. Without loss of generality, take $\|T\| = 1$.

From the Hahn–Banach theorem, choose $\phi \in \mathsf{M}_n^*$, $\|\phi\| = 1$, such that

$$\phi(T) = 1, \quad \phi(I) = 0.$$

This functional has the form

$$\phi(A) = \sum_{i=1}^{k} \lambda_i \langle A\xi_i, \eta_i \rangle, \qquad A \in \mathsf{M}_n$$

for unit vectors $\xi_i, \eta_i \in \mathbb{C}^n$, $\lambda_i > 0$, $\sum_{i=1}^{k} \lambda_i = 1$, since the unit ball of M_n^* is the convex hull of vector functionals $\langle \cdot\xi, \eta \rangle$, $\|\xi\| = \|\eta\| = 1$. The relations

$$1 = \phi(T) = \|T\|$$

force $T\xi_i = \eta_i$, $1 \leq i \leq k$, and so $T^*\eta_i = \xi_i$. Thus

$$0 = \phi(I) = \sum_{i=1}^{k} \lambda_i \langle \xi_i, \eta_i \rangle = \sum_{i=1}^{k} \lambda_i \langle \xi_i, T\xi_i \rangle$$

so

$$\sum_{i=1}^{k} \lambda_i \langle T\xi_i, \xi_i \rangle = 0.$$

Let $H = \mathrm{span}\{\xi_1, \ldots, \xi_k\}$ and let P be the projection of $\mathbb{C}^n$ onto H. Then

$$\sum_{i=1}^{k} \lambda_i \langle PTP\xi_i, \xi_i \rangle = 0$$

so 0 is in the numerical range of $PTP \in B(H)$ [BD1]. Thus there is a unit vector $\xi \in H$ such that

$$\langle PTP\xi, \xi \rangle = 0.$$

It follows that $\langle T\xi, \xi \rangle = 0$. From above

$$T^*T\xi_i = T^*\eta_i = \xi_i, \qquad 1 \leq i \leq k$$

and so T is an isometry on H. Thus $\|T\xi\| = 1$, and so we may choose a unitary $U \in \mathbf{M}_n$ such that

$$UT\xi = T\xi, \quad U\xi = -\xi.$$

Then

$$\|(UT - TU)\xi\| = \|T\xi + T\xi\| = 2$$

and so

$$\|\delta\| \geq \|\delta(U)\| \geq 2 = 2\|T\|.$$

The reverse inequality is obvious.

2.5.3 Lemma. *Let $T \in B(H)$ and let δ be the inner derivation implemented by T. Then*

$$\|\delta\| = 2 \inf_{\lambda \in \mathbb{C}} \|T - \lambda I\| = 2 \, \mathrm{dist}(T, \mathbb{C}I).$$

Proof. As in Lemma 2.5.2, we may assume that

$$1 = \|T\| = \mathrm{dist}(T, \mathbb{C}I)$$

and then prove that $\|\delta\| \geq 2$.

Fix $\varepsilon > 0$, let $D = \{\lambda \in \mathbb{C} \colon |\lambda| \leq 2\}$ and choose $\lambda_1, \ldots, \lambda_n \in D$ such that

$$\inf_{1 \leq i \leq n} |\lambda_i - \lambda| \leq \varepsilon, \qquad \lambda \in D.$$

Then choose unit vectors $\xi_1, \ldots, \xi_n \in H$ such that

$$\|(\lambda_i I - T)\xi_i\| \geq 1 - \varepsilon, \qquad 1 \leq i \leq n.$$

Let $K = \text{span}\{\xi_i, T\xi_i \colon 1 \le i \le n\}$ and let P be the projection onto K. Then, in $B(K)$,

$$\|\lambda_i I - PTP\| \ge \|(\lambda_i I - PTP)\xi_i\| \ge 1 - \varepsilon, \qquad 1 \le i \le n.$$

If $\lambda \in D$, choose $\lambda_i \in D$ such that $|\lambda - \lambda_i| \le \varepsilon$. Then it is clear that $\|\lambda I - PTP\| \ge 1 - 2\varepsilon$, while if $|\lambda| > 2$ then $\|\lambda I - PTP\| > 1$. Thus

$$\text{dist}(PTP, \mathbb{C}I) \ge 1 - 2\varepsilon.$$

Let δ_1 be the derivation on $B(K)$ implemented by PTP. By Lemma 2.5.2,

$$\|\delta_1\| \ge 2 - 4\varepsilon$$

and so there exists a unitary $U \in B(K)$ and a unit vector $\xi \in K$ such that

$$\|(UPTP - PTPU)\xi\| \ge 2 - 4\varepsilon.$$

Let $V \in B(H)$ be a unitary whose restriction to K is U. Then, from above,

$$\|PT\xi\| = \|UPTP\xi\| \ge 1 - 4\varepsilon.$$

Since

$$T\xi = PT\xi + P^{\perp}T\xi,$$

it follows that

$$\begin{aligned} \|P^{\perp}T\xi\|^2 = \|T\xi\|^2 - \|PT\xi\|^2 &\le 1 - (1 - 4\varepsilon)^2 \\ &= 8\varepsilon - 16\varepsilon^2 \le 16\varepsilon \end{aligned}$$

so that $\|P^{\perp}T\xi\| \le 4\varepsilon^{1/2}$. Then

$$\begin{aligned} \|VT\xi - TV\xi\| = \|VPT\xi + VP^{\perp}T\xi - TV\xi\| \\ \ge \|VPT\xi - TV\xi\| - 4\varepsilon^{1/2} \\ = \|UPTP\xi - PTPU\xi\| - 4\varepsilon^{1/2} \\ \ge 2 - 4\varepsilon - 4\varepsilon^{1/2}. \end{aligned}$$

Since $\varepsilon > 0$ was arbitrary, we conclude that $\|\delta\| \ge 2$.

2.5.4 Theorem. *Let $\mathcal{M}$ be a von Neumann algebra with centre $\mathcal{Z}$ and let $m_0 \in \mathcal{M}$ implement a derivation δ. Then*

$$\|\delta\| = 2 \, \text{dist}(m_0, \mathcal{Z}).$$

Proof. Inequality in one direction is clear. The distance of m_0 to $\mathcal{Z}$ is attained since $\mathcal{Z}$ is a von Neumann algebra, so we may assume that

$$\|m_0\| = \text{dist}(m_0, \mathcal{Z}).$$

We may then scale m_0 so that $\|m_0\| = 1$, and we will show that $\|\delta\| \geq 2$.

The centre is an abelian C^*-algebra, so may be regarded as $C(\Omega)$ for some compact Hausdorff space Ω. Each $\omega \in \Omega$ defines a maximal ideal $\mathcal{I}_\omega \subseteq \mathcal{Z}$, which in turn generates a norm closed ideal $\mathcal{J}_\omega \subseteq \mathcal{M}$. Each of these ideals $\mathcal{J}_\omega$ is the kernel of an irreducible representation $\pi_\omega \colon \mathcal{M} \to B(H_\omega)$ [Hal]. We note that each $\mathcal{J}_\omega$ has an approximate identity lying in $\mathcal{I}_\omega$.

We first show that $\bigcap_{\omega \in \Omega} \mathcal{J}_\omega = 0$. If not, then there is a non-zero element m in this intersection, so we may choose a pure state ϕ on $\mathcal{M}$ so that $\phi(m) \neq 0$. Let $\pi \colon \mathcal{M} \to B(H_\pi)$ be the irreducible representation associated to ϕ from the GNS construction. By irreducibility, π maps $\mathcal{Z}$ into $\mathbb{C}I$ and so $\ker \pi$ contains $\mathcal{I}_\omega$ for some $\omega \in \Omega$ and thus also contains $\mathcal{J}_\omega$. Then $\pi(m) = 0$. However, there is a vector $\xi \in H_\pi$ such that

$$\phi(x) = \langle \pi(x)\xi, \xi \rangle, \qquad x \in \mathcal{M},$$

and so $\phi(m) = 0$, a contradiction.

From this calculation, we see that $\bigoplus_{\omega \in \Omega} \pi_\omega$ is a faithful representation of $\mathcal{M}$, and so

$$\|x\| = \sup_{\omega \in \Omega} \|\pi_\omega(x)\| = \sup_{\omega \in \Omega} \|x + \mathcal{J}_\omega\|, \qquad x \in \mathcal{M}.$$

Now suppose that $\|\delta\| < 2$, so that we may choose $\varepsilon > 0$ with $\|\delta\| \leq 2 - 8\varepsilon$. By Lemma 2.2.3, δ induces derivations

$$\delta_\omega \colon \pi_\omega(\mathcal{M}) \to \pi_\omega(\mathcal{M}), \qquad \omega \in \Omega,$$

since $\pi_\omega(\mathcal{M})$ is isomorphic to $\mathcal{M}/\mathcal{J}_\omega$. Each δ_ω is implemented by $\pi_\omega(m_0)$ and we may regard δ_ω as a derivation on $B(H_\omega)$ since $\pi_\omega(\mathcal{M})$ is strongly dense. The Kaplansky density theorem shows that $\|\delta_\omega\| \leq 2 - 8\varepsilon$ as a derivation on $B(H_\omega)$ and so, by Lemma 2.5.3, there exists $\lambda_\omega \in \mathbb{C}$ such that $\|\pi_\omega(m_0) - \lambda_\omega I\| \leq 1 - 4\varepsilon$. For a fixed but arbitrary $\omega \in \Omega$, there exists $j_\omega \in \mathcal{J}_\omega$ such that

$$\|m_0 - \lambda_\omega 1 - j_\omega\| \leq 1 - 3\varepsilon.$$

Then choose $z \in \mathcal{I}_\omega$ such that $\|zj_\omega - j_\omega\| \leq \varepsilon$, and let U_ω be the open neighbourhood of ω defined by

$$\{\mu \in \Omega : |z(\mu)| < \varepsilon\|j_\omega\|^{-1}\}.$$

Then $(z - z(\mu)1)j_\omega \in \mathcal{J}_\mu$ and

$$\|m_0 - \lambda_\omega 1 - (z - z(\mu)1)j_\omega\| \leq \|m_0 - \lambda_\omega 1 - zj_\omega\| + \|z(\mu)j_\omega\|$$
$$\leq 1 - 2\varepsilon + \varepsilon$$
$$= 1 - \varepsilon$$

if $\mu \in U_\omega$. Thus

$$\|\pi_\mu(m_0 - \lambda_\omega 1)\| \leq 1 - \varepsilon$$

for $\mu \in U_\omega$. By compactness, select a finite subcover $U_{\omega_1}, \ldots, U_{\omega_n}$ of Ω and let $\{z_1, \ldots, z_n\}$ be a partition of unity:

$$0 \leq z_i \leq 1, \quad \sum_{i=1}^{n} z_i = 1, \quad z_i \in \mathcal{Z},$$

and each z_i is supported on U_{ω_i}. Now consider $m_0 - \sum_{i=1}^{n} \lambda_{\omega_i} z_i$. For a fixed but arbitrary $\mu \in \Omega$ we may assume that

$$\mu \in U_{\omega_1}, \ldots, U_{\omega_k}, \quad \mu \notin U_{\omega_{k+1}}, \ldots, U_{\omega_n},$$

renumbering if necessary. Then

$$\pi_\mu \left(\sum_{i=1}^{k} z_i \right) = I, \quad \pi_\mu(z_i) = 0, \quad i \geq k + 1.$$

Thus

$$\left\| \pi_\mu \left(m_0 - \sum_{i=1}^{n} \lambda_{\omega_i} z_i \right) \right\| = \left\| \sum_{i=1}^{k} z_i(\mu) \pi_\mu(m_0 - \lambda_{\omega_i} 1) \right\|$$
$$\leq 1 - \varepsilon$$

since $\sum_{i=1}^{k} z_i(\mu) = 1$ and $\mu \in \bigcap_{i=1}^{k} U_{\omega_i}$. We have now shown that

$$\left\| m_0 - \sum_{i=1}^{n} \lambda_{\omega_i} z_i \right\| \leq 1 - \varepsilon < 1,$$

which contradicts our original assumptions on m_0. Thus $\|\delta\| \geq 2$.

2.5.5 Corollary. *Let $\delta \colon \mathcal{M} \to \mathcal{M}$ be a derivation on a von Neumann algebra $\mathcal{M}$. Then δ is implemented by an element $m_0 \in \mathcal{M}$ of norm $\|\delta\|/2$.*

Proof. This is just a reformulation of the preceding theorem.

2.6 Notes and Remarks

In this chapter we have restricted attention to bounded derivations. The reader who is interested in unbounded derivations should consult the book by Sakai [S3] which also contains much of the theory presented here, although using different methods.

The main result, Theorem 2.5.1, is the celebrated Kadison–Sakai theorem that all derivations on von Neumann algebras are inner [Ka2, S1]. The first few sections are essentially taken from these two papers, although for the step from knowing that derivations are spatial to showing that they are inner, we draw on recent results on completely bounded module maps [ChS5]. The minimal norm estimates of Lemma 2.5.3 and Theorem 2.5.4 are due respectively to Stampfli [St] and Zsidó [Zs] (see [Ga] for the separable Hilbert space case).

We have confined the discussion to derivations of $\mathcal{M}$ into itself where the results are definitive. The situation is unclear for derivations into $B(H)$ (or more general modules). It is known that $\delta \colon \mathcal{M} \to B(H)$ is spatially implemented if and only if δ is completely bounded [Ch5], and in certain circumstances complete boundedness can be verified [Ch5] (see also [Ha2]). In all known cases, derivations are spatially implemented, and the only remaining situation to consider is a type II_1 von Neumann algebra with a type II_∞ commutant.

3

Averaging in Continuous and Normal Cohomology

3.1 Introduction

This chapter contains the basic averaging techniques and the method for lifting a continuous multilinear map to the weak closure of the algebra. The averaging first leads to the cocycles being replaced by cocycles that are module maps over hyperfinite (= injective) subalgebras, and then to completely bounded maps in suitable situations (see Section 6.2). The weak lifting results are used to prove that continuous and normal cohomology are equal for dual normal modules over von Neumann algebras. Both the above methods will be used on continuous and completely bounded multilinear maps, with the proofs for the latter being just a minor modification of the continuous case developed by Johnson, Kadison and Ringrose.

This introduction will contain a discussion of the averaging techniques and the normal operator lifting method, and the rest of the chapter will be devoted to the details. The averaging techniques over C^*-algebras are in Section 3.2 and the normal operator lifting methods are in Section 3.3, as these require the averaging technique at one critical point. The two sections come together in Section 3.4 in averaging over a hyperfinite von Neumann subalgebra.

The averaging technique used is basically just to integrate over the compact unitary group of a finite dimensional C^*-algebra. Taking suitable weak limits as the finite dimensional algebras increase in size leads to averages that are essentially over infinite dimensional algebras. A similar route is to use an amenable group of unitary operators and the (right) invariant mean to average. This is the path taken in [Ri3,6] which will be followed here. This chapter is based on Ringrose's survey [Ri6]. Though the technical lemmas are worded in a different way and are given in greater generality (e.g. the completely bounded case), the method and approach is that in his notes. A more abstract approach would be to use the amenability of the algebra being averaged over directly as in [JKR]. This method is logically direct but entails constructing various dual modules and derivations to which the amenability hypothesis may be applied. These constructions are carried out in detail on pp. 84–87 of [JKR] in the continuous case. However they seemed less intuitive to us than averaging, and technical constructions using operator spaces [BP], [ER1] are required in the completely bounded category. There is further discussion of this point and of a more axiomatic approach in the Notes and Remarks (Section 3.5).

The main results in this chapter are Theorems 3.3.1 and 3.4.1, which may be summarized as follows. See Sections 0.2 and 0.6 for definitions and notation.

3.1.1 Theorem. *If $\mathcal{N}$ is a hyperfinite von Neumann subalgebra of a von Neumann algebra $\mathcal{M}$ and if $\mathcal{V}$ is a dual normal $\mathcal{M}$-module, then*

$$H^n(\mathcal{M}, \mathcal{V}) \cong H_w^n(\mathcal{M}, \mathcal{V}) \cong H_w^n(\mathcal{M}, \mathcal{V} : /\mathcal{N})$$

and

$$H_{cb}^n(\mathcal{M}, \mathcal{V}) \cong H_{wcb}^n(\mathcal{M}, \mathcal{V}) \cong H_{wcb}^n(\mathcal{M}, \mathcal{V} : /\mathcal{N}),$$

where all the isomorphisms are induced naturally by the corresponding embeddings of the space of maps $\mathcal{L}^n(\mathcal{M}, \mathcal{V})$.

Along the way to proving this result, a number of technical lemmas on C^*-algebra and von Neumann algebra cohomology are proved. These lead to equality of the continuous cohomology of a represented C^*-algebra over a dual normal module with the cohomology of its ultraweak closure (Theorem 3.3.1). To emphasize the categorical nature of all the calculations involved, maps between the complexes are introduced and used to prove the various isomorphisms. The most important of these maps are listed below and then there is a brief description of their use. For the precise properties of these maps and their definition it is essential for the reader to turn to the lemmas referred to below.

$$
\begin{aligned}
&K_n\colon \mathcal{L}^n(\mathcal{A}, \mathcal{V}) \to \mathcal{L}^{n-1}(\mathcal{A}, \mathcal{V}) && (3.2.4)\\
&L_n\colon \mathcal{L}_w^n(\mathcal{M}, \mathcal{V}) \to \mathcal{L}_w^{n-1}(\mathcal{M}, \mathcal{V}) && (3.4.2)\\
&Q_n\colon \mathcal{L}^n(\mathcal{A}, \mathcal{V}) \to \mathcal{L}^n(\mathcal{A}, \mathcal{V} : /\mathcal{B}) && (3.2.6)\\
&P_n\colon \mathcal{L}_w^n(\mathcal{M}, \mathcal{V}) \to \mathcal{L}_w^n(\mathcal{M}, \mathcal{V} : /\mathcal{B}) && (3.4.3)\\
&J_n\colon \mathcal{L}^n(\mathcal{A}, \mathcal{V}) \to \mathcal{L}^{n-1}(\mathcal{A}, \mathcal{V}) && (3.2.4)\\
&S_n T_n\colon \mathcal{L}^n(\mathcal{A}, \mathcal{V}) \to \mathcal{L}_w^n(\bar{\mathcal{A}}, \mathcal{V}) && (3.3.4).
\end{aligned}
$$

The pairs of operators (K_n, Q_n) and (L_n, P_n) play analogous roles in the continuous and normal cases, with (K_n, Q_n) leading to

$$H^n(\mathcal{A}, \mathcal{V}) \cong H^n(\mathcal{A}, \mathcal{V} : /\mathcal{B}) \qquad (3.2.7)$$

and (L_n, P_n) to

$$H_w^n(\mathcal{M}, \mathcal{V}) \cong H_w^n(\mathcal{M}, \mathcal{V} : /\mathcal{N}) \qquad (3.4.1).$$

The same operators also handle the completely bounded situation. The pair $(J_n, S_n T_n)$ is used to prove

$$H^n(\mathcal{M}, \mathcal{V}) \cong H_w^n(\mathcal{M}, \mathcal{V}) \qquad (3.3.1)$$

which is the purely topological stage of our equalities. Where appropriate, this pair $(J_n, S_n T_n)$ also deals with module maps.

3.2 Averaging in Continuous Cohomology

The main result of this section is Lemma 3.2.4, which shows that a co-cycle may be modified by a coboundary to be a module map with respect to a subalgebra generated by a compact or amenable subgroup. There is a discussion of averaging in Remark 3.2.2. The following little algebraic lemma is useful at several points in these notes to observe that a cocycle zero on a subalgebra is a module map over the subalgebra. Except in the proof of Lemma 3.2.4 it is only used when $k = n$.

3.2.1 Lemma. *Let $\mathcal{B}$ be a unital subalgebra of a unital C^*-algebra $\mathcal{A}$, let $\mathcal{V}$ be a Banach $\mathcal{A}$-module, and let ϕ be in $\mathcal{L}^n(\mathcal{A}, \mathcal{V})$ with $\partial\phi = 0$. Let $1 \leq k \leq n$. Then*

(1) $$\phi(x_1, \ldots, x_{j-1}, b, x_{j+1}, \ldots, x_n) = 0$$

for all $b \in \mathcal{B}$, $x_1, \ldots, x_n \in \mathcal{A}$ and $1 \leq j \leq k$ if and only if the following three conditions hold:

(2) $$\phi(x_1, \ldots, x_{j-1}, 1, x_{j+1}, \ldots, x_n) = 0,$$
(3) $$\phi(bx_1, x_2, \ldots, x_n) = b\phi(x_1, \ldots, x_n)$$

and

(4) $$\phi(x_1, \ldots, x_j b, x_{j+1}, \ldots, x_n) = \phi(x_1, \ldots, x_j, bx_{j+1}, \ldots, x_n)$$

for all $b \in \mathcal{B}$, $x_1, \ldots, x_n \in \mathcal{A}$ and $1 \leq j \leq k$.

If $k = n$, then (1) implies that

(5) $$\phi(x_1, \ldots, x_n b) = \phi(x_1, \ldots, x_n)b$$

for all $x_1, \ldots, x_n \in \mathcal{A}$ and $b \in \mathcal{B}$.

Proof. Clearly (2) and (4) imply (1). Suppose that (1) holds. This implies that all but two of the terms in the sum defining $\partial\phi$ are zero in

$$\partial\phi(x_1, \ldots, x_j, b, x_{j+1}, \ldots, x_n) = 0.$$

These two terms are of opposite sign and are

$$\phi(x_1, \ldots, x_j b, x_{j+1}, \ldots, x_n) \quad \text{and} \quad \phi(x_1, \ldots, x_j, bx_{j+1}, \ldots, x_n).$$

This proves (4). Equations (2) and (5) are obtained using the same technique by letting $\partial \phi = 0$ act respectively on

$$(b, x_1, \ldots, x_n) \quad \text{and} \quad (x_1, \ldots, x_n, b).$$

3.2.2 Remarks and Definitions. One of the standard techniques in the study of cohomology and of derivations is to average over a suitable group. For algebraic maps the average can only be taken over a finite group, and the lack of limiting processes is restrictive. For continuous maps associated with Banach algebras and modules, an average over the Haar measure of a compact group gives maps of the same type. The reason is that these integrals are norm convergent, and do not involve weak or weak*-limits, so even averaging normal multilinear maps gives maps with the same continuity properties. However, subsequently it is essential to average over much larger subalgebras and their generating amenable groups of unitaries. This is usually done by considering a right (or left) invariant mean μ on a suitable amenable group and using weak* (or weak) limits. This is the method used below in Lemma 3.2.4 (b), which lies behind many of the theorems in the rest of these notes. However, in the von Neumann situation, the amenable groups used can be chosen to be unions of increasing nets of compact groups of unitaries. This enables us to take a weak*-limit over the averages obtained on this net of compact groups and to use Lemma 3.2.4 (a). Of course, the final limit is still a weak* limit so the difficulty of averaging normal maps and obtaining normal ones remains. The normal cocycles and averaging are postponed until Section 3.4, and are handled by Lemma 3.2.4 and results from Section 3.3 which modify the continuous averages to be normal by subtracting a suitable coboundary.

Let $\mathcal{U}$ be a locally compact group and let $C(\mathcal{U})$ denote the Banach space of all continuous bounded complex-valued functions on $\mathcal{U}$. A *right invariant mean* μ on $\mathcal{U}$ is a continuous linear functional of norm 1 on $C(\mathcal{U})$ with $\mu(1) = 1$ such that

$$\mu(f_w) = \mu(f) \text{ for all } f \in C(\mathcal{U}) \text{ and all } w \in \mathcal{U},$$

where $f_w(u) = f(uw)$ for all $u \in \mathcal{U}$. These conditions imply that $\mu(f) \geq 0$ if $f \geq 0$ in $C(\mathcal{U})$. A locally compact group $\mathcal{U}$ is said to be *amenable* if such a right invariant mean μ exists [Gre, P1, Pe2]. It helps to think of μ as a "measure" on $\mathcal{U}$ that satisfies

$$\int f(uw)d\mu(u) = \int f(u)d\mu(u)$$

for all $f \in C(\mathcal{U})$ and all $w \in \mathcal{U}$. This point of view makes the transition from the proof of part (a) to part (b) of Lemma 3.2.4 a formality. However, care is needed as such a measure is not countably additive. If $\mathcal{V}$ is a dual Banach space with predual $\mathcal{V}_*$ and Φ is a continuous function from $\mathcal{U}$ into $\mathcal{V}$, then the invariant mean

$$\int \Phi(u)d\mu(u)$$

of Φ is defined by

$$\left\langle \int \Phi(u)d\mu(u), x \right\rangle = \int \langle \Phi(u), x \rangle d\mu(u)$$

for all $x \in \mathcal{V}_*$. Note that the right-hand side of this equation induces a continuous linear functional on $\mathcal{V}_*$, so defining the element $\int \Phi(u)d\mu(u)$ in $\mathcal{V}$. Using the right invariance of the invariant mean on $\mathcal{U}$ and the definition yields

$$\int \Phi(uv)d\mu(u) = \int \Phi(u)d\mu(u)$$

for all $v \in \mathcal{U}$.

In these notes, we are only interested in averaging over suitable subgroups of the unitary group of a hyperfinite von Neumann alegbra $\mathcal{M}$, which is the ultraweak closure of an increasing family $\mathcal{N}_\lambda$ of finite dimensional subalgebras. Unfortunately, the restriction of the standard topologies on $\mathcal{M}$ to a subgroup $\mathcal{U}$ of the unitary group gives in general topological groups, but not necessarily locally compact groups. There are two ways to circumvent this technical difficulty. The first is to average over the compact unitary groups of the $\mathcal{N}_\lambda$'s and take ultraweak limits over λ. The second is to recall, from Lemma 2.4.1, that each $\mathcal{N}_\lambda$ contains a finite group G_λ which generates it. The G_λ's may be chosen to be nested, and so $\mathcal{U} = \bigcup_\lambda G_\lambda$ with the discrete topology is a locally compact amenable group of unitaries generating $\mathcal{M}$. In the subsequent lemmas we do not assume that the topologies on the amenable subgroups $\mathcal{U}$ are inherited from their ambient C^*-algebras or von Neumann algebras. The average by an invariant mean μ will be used for only a few types of functions Φ derived from multilinear maps.

3.2.3 Lemma. *Let $\mathcal{A}$ be a unital C^*-subalgebra of $B(H)$, let $\mathcal{V}$ be a dual normal $\mathcal{A}$-module, let $\mathcal{U}$ be an amenable group of unitaries contained in $\mathcal{A}$ and let $\mathcal{B}$ be the C^*-algebra generated by $\mathcal{U}$. If $\phi \in \mathcal{L}^n(\mathcal{A}, \mathcal{V})$, then each of the following maps is in $\mathcal{L}^{n-1}(\mathcal{A}, \mathcal{V})$ or $\mathcal{L}^n(\mathcal{A}, \mathcal{V})$ (depending on the number of variables), and has norm no greater than $\|\phi\|$, even when the subscript "cb" is adjoined to $\mathcal{L}$ on both sides and to the norm.*

(1) $(x_1, \ldots, x_{n-1}) \mapsto \int u^*\phi(u, x_1, \ldots, x_{n-1})d\mu(u),$

(2) $(x_1, \ldots, x_{n-1}) \mapsto \int \phi(x_1, \ldots, x_{k-1}, x_k u^*, u, x_{k+1}, \ldots, x_{n-1}) d\mu(u),$

(3) $(x_1, \ldots, x_n) \mapsto \int u^* \phi(u x_1, x_2, \ldots, x_n) d\mu(u),$

(4) $(x_1, \ldots, x_n) \mapsto \int \phi(x_1, \ldots, x_{k-1}, x_k u^*, u x_{k+1}, \ldots, x_n) d\mu(u),$

(5) $(x_1, \ldots, x_n) \mapsto \int \phi(x_1, \ldots, x_{n-1}, x_n u^*) u \, d\mu(u),$

where $1 < k \le n$. The maps in (3), (4) and (5) each commute with elements of $\mathcal{B}$ across the averaged position.

Proof. The map in (4) will be discussed in detail; the others all follow by similar calculations. Directly from the definition, it follows that the map

$$\psi(x_1, \ldots, x_n) \equiv \int \phi(x_1, \ldots, x_{k-1}, x_k u^*, u x_{k+1}, \ldots, x_n) d\mu(u)$$

is n-linear, so lies in $\mathcal{L}^n(\mathcal{A}, \mathcal{V})$. Further, $\|\psi\| \le \|\phi\|$ in the continuous case and

$$\psi(x_1, \ldots, x_k v, x_{k+1}, \ldots, x_n)$$
$$= \int \phi(x_1, \ldots, x_k(uv^*)^*, (uv^*) v x_{k+1}, \ldots, x_n) d\mu(u)$$
$$= \int \phi(x_1, \ldots, x_k u^*, uv x_{k+1}, \ldots, x_n) d\mu(u)$$
$$= \psi(x_1, \ldots, x_k, v x_{k+1}, \ldots, x_n)$$

for all $v \in \mathcal{U}$ by the right invariance of μ. The continuity of ψ and the density of the linear span of $\mathcal{U}$ in $\mathcal{B}$ implies the module property of ψ from the kth to $(k+1)^{\text{th}}$ variables. The m-fold amplification ψ_m of ψ (Section 1.5) is a linear definition without change of order of the basic variables. Writing the definition of m-fold amplification out in full and letting an element of $\mathbf{M}_m(\mathcal{V}_*)$ act on $\psi_m(x_1, \ldots, x_n)$ by the usual tracial action on $\mathbf{M}_m(\mathcal{V})$ gives

$$\psi_m(x_1, \ldots, x_n) = \int \phi_m(x_1, \ldots, x_{k-1}, x_k(u \otimes I)^*, (u \otimes I) x_{k+1}, \ldots, x_n) d\mu(u)$$

for all $x_1, \ldots, x_n \in \mathbf{M}_m(\mathcal{A})$. This is just the observation that amplifications and averages commute. From this equation it follows that $\|\psi_m\| \le \|\phi_m\|$ for all m, so that $\|\psi\|_{cb} \le \|\phi\|_{cb}$ in the completely bounded situation.

3.2.4 Lemma. (a) Let $\mathcal{B}$ be a finite dimensional unital $*$-subalgebra of a unital C^*-algebra $\mathcal{A}$ and let $\mathcal{V}$ be a dual normal Banach $\mathcal{A}$-module. There is a continuous linear map $K_n : \mathcal{L}^n(\mathcal{A}, \mathcal{V}) \to \mathcal{L}^{n-1}(\mathcal{A}, \mathcal{V})$ (depending on $\mathcal{B}$) such that ϕ in $\mathcal{L}^n(\mathcal{A}, \mathcal{V})$ with $\partial \phi = 0$ implies that $\phi - \partial(K_n \phi)$ is a $\mathcal{B}$-module map. Further the subscripts "w", "cb", or "wcb" may be adjoined to $\mathcal{L}$ throughout, and in each case

$$\|K_n\| \le ((n+2)^n - 1)/(n+1).$$

If $C \subseteq A$ *commutes with* B, *then* K_n *maps* $\mathcal{L}^n(A, V: /C)$ *into* $\mathcal{L}^{n-1}(A, V: /C)$.

(b) Let B *be the* C^*-*algebra generated by an amenable group* $\mathcal{U}$ *of unitaries in a unital* C^*-*subalgebra* A *of* $B(H)$, *and let* V *be a dual normal* A-*module. There is a continuous linear map* K_n: $\mathcal{L}^n(A, V) \to \mathcal{L}^{n-1}(A, V)$ *(depending on* $\mathcal{U}$*) such that* ϕ *in* $\mathcal{L}^n(A, V)$ *with* $\partial\phi = 0$ *implies that* $\phi - \partial(K_n\phi)$ *is* B-*multimodular. Further, the subscript "cb" may be adjoined to* $\mathcal{L}$ *throughout, and in both cases*

$$\|K_n\| \leq ((n+2)^n - 1)/(n+1).$$

Proof. The proof is essentially the same in both cases with the average in the amenable case having always to be defined in terms of the weak*-topology on V. Case (a) will be proved initially, then there is a brief discussion of (b). Let $\mathcal{U}$ be the compact unitary group of B; μ will be Haar measure on $\mathcal{U}$ normalized by $\mu(\mathcal{U}) = 1$ and $\int \ldots d\mu(u)$ will denote the integral over $\mathcal{U}$ with respect to μ. Throughout let ϕ be an n-cocycle, so that $\partial\phi = 0$.

Maps $J_1, \ldots, J_n$ from $\mathcal{L}^n(A, V)$ into $\mathcal{L}^{n-1}(A, V)$ are constructed by induction such that

$$(\phi - \partial J_k\phi)(x_1, \ldots, x_n) = 0$$

if any one of $x_1, \ldots, x_k$ is in B, and such that $\|J_k\| \leq ((n+2)^k - 1)/(n+1)$ for $k = 1, \ldots, n$. Then J_n is the required map K_n.

Define

$$(1) \qquad (J_1\psi)(x_1, \ldots, x_{n-1}) = \int u^*\psi(u, x_1, \ldots, x_{n-1})d\mu(u)$$

for all $\psi \in \mathcal{L}^n(A, V)$ and $x_1, \ldots, x_{n-1} \in A$. Clearly J_1 is a continuous linear map from $\mathcal{L}^n(A, V)$ into $\mathcal{L}^{n-1}(A, V)$ with $\|J_1\| \leq 1$. Now

$$(2) \quad x_1 u^*\phi(u, x_2, \ldots, x_n) + \sum_{j=1}^{n}(-1)^j u^*\phi(u, x_1, \ldots, x_j x_{j+1}, \ldots, x_n)$$

$$+ (-1)^n u^*\phi(u, x_1, \ldots, x_{n-1})x_n$$

$$= x_1 u^*\phi(u, x_2, \ldots, x_n) - u^*\partial\phi(u, x_1, \ldots, x_n)$$

$$+ u^*\{u\phi(x_1, \ldots, x_n) - \phi(ux_1, \ldots, x_n)\}$$

$$\text{by definition of } \partial\phi(u, x_1, \ldots, x_n)$$

$$= \phi(x_1, \ldots, x_n) + x_1 u^*\phi(u, x_2, \ldots, x_n)$$

$$- u^*\phi(ux_1, \ldots, x_n).$$

because $\partial\phi = 0$.

If $x_1 \in \mathcal{U}$, then averaging equation (2) over $\mathcal{U}$ gives

$$\partial J_1 \phi(x_1, \ldots, x_n) = x_1 J_1 \phi(x_2, \ldots, x_n)$$

$$+ \sum_{j=1}^{n-1} (-1)^j J_1 \phi(x_1, \ldots, x_j x_{j+1}, \ldots, x_n)$$

$$+ (-1)^n J_1 \phi(x_1, \ldots, x_{n-1}) x_n$$

$$= \phi(x_1, \ldots, x_n) + x_1 \int u^* \phi(u, \ldots, x_n) d\mu(u)$$

$$- x_1 \int (ux_1)^* \phi(ux_1, \ldots, x_n) d\mu(u)$$

$$= \phi(x_1, \ldots, x_n)$$

by the right invariance of μ. The continuity and linearity of $\phi - \partial J_1 \phi$ in the first variable implies that

$$(\phi - \partial J_1 \phi)(x_1, \ldots, x_n) = 0$$

if $x_1 \in \mathcal{B}$.

Suppose that J_k has been constructed so that

(3) $$(\phi - \partial J_k \phi)(x_1, \ldots, x_n) = 0$$

if any one of $x_1, \ldots, x_k \in \mathcal{B}$ and that

(4) $$\|J_k\| \leq ((n+2)^k - 1)/(n+1).$$

Let

(5) $$\chi = \phi - \partial J_k \phi.$$

Define $G: \mathcal{L}^n(\mathcal{A}, \mathcal{V}) \to \mathcal{L}^{n-1}(\mathcal{A}, \mathcal{V})$ by
(6)

$$(G\psi)(x_1, \ldots, x_{n-1}) = \int \psi(x_1, \ldots, x_{k-1}, x_k u^*, u, x_{k+1}, \ldots, x_{n-1}) d\mu(u)$$

for all $x_1, \ldots, x_{n-1} \in \mathcal{A}$. Clearly G is a continuous linear map with $\|G\| \leq 1$.

Now by (5) and (6)

(7) $$(G\chi)(x_1, \ldots, x_n) = 0$$

if any one of $x_1, \ldots, x_k$ is in $\mathcal{B}$, because $G\chi$ is an average of maps that are zero by (3). From (3), the equation $\partial\chi = 0$, and Lemma 3.2.1 it follows that

$$b\chi(x_1, \ldots, x_k u^*, u, x_{k+1}, \ldots, x_{n-1}) =$$

$$\chi(bx_1, \ldots, x_k u^*, u, x_{k+1}, \ldots, x_{n-1}),$$

$$\chi(x_1, \ldots, x_{j-1} b, x_j, \ldots, x_k u^*, u, \ldots, x_{n-1}) =$$

$$\chi(x_1, \ldots, x_{j-1}, bx_j, \ldots, x_k u^*, u, \ldots, x_{n-1})$$

for $1 < j < k - 1$, and

$$\chi(x_1, \ldots, x_{k-1}b, x_k u^*, u, \ldots, x_{n-1}) = \chi(x_1, \ldots, x_{k-1}, bx_k u^*, u, \ldots, x_{n-1})$$

for all $b \in \mathcal{B}$ and $x_1 \ldots, x_{n-1} \in \mathcal{A}$. Averaging these equations over $u \in \mathcal{U}$ yields

$$(8) \qquad bG\chi(x_1, \ldots, x_{n-1}) = G\chi(bx_1, \ldots, x_{n-1})$$

and

$$(9) \quad G\chi(x_1, \ldots, x_{j-1}b, x_j, \ldots, x_{n-1}) = G\chi(x_1, \ldots, x_{j-1}, bx_j, \ldots, x_{n-1}).$$

for $1 < j \le k$ and all $b \in \mathcal{B}$. Define

$$(10) \qquad J_{k+1} = J_k + (-1)^k G(I - \partial J_k)$$

so that

$$(11) \qquad J_{k+1}\phi = J_k\phi + (-1)^k G_\chi.$$

Then J_{k+1} is a continuous linear map from $\mathcal{L}^n(\mathcal{A}, \mathcal{V})$ into $\mathcal{L}^{n-1}(\mathcal{A}, \mathcal{V})$.
The next stage in the proof is to show that

$$(\phi - \partial J_{k+1}\phi)(x_1, \ldots, x_n)$$

is zero if any one of the first k variables is in $\mathcal{B}$; the $(k+1)^{\text{th}}$ variable is then tackled separately. Let $1 \le j \le k$ and let $x_j \in \mathcal{B}$. By (8), (9) and the definition of ∂ it follows that all the terms in $\partial G\chi(x_1, \ldots, x_n)$ are zero except for the jth and $(j+1)^{\text{th}}$ terms. These are of opposite sign, being successive terms in ∂, and are equal by equations (8) (for $j = 1$) and (9) (for $1 < j \le k$). Hence $\partial G\chi(x_1, \ldots, x_n) = 0$. Combining this with (11) and the inductive assumption (3), we obtain

$$(12) \qquad (\phi - \partial J_{j+1}\phi)(x_1, \ldots, x_n) = 0.$$

We now turn to the $(k+1)^{\text{th}}$ variable. To prove (12) when x_{k+1} is in $\mathcal{B}$ it is sufficient to prove it when $x_{k+1} \in \mathcal{U}$, as $\mathcal{B}$ is the closed linear span of $\mathcal{U}$. The first k terms in the expansion of $\partial G\chi(x_1, \ldots, x_n)$ are zero by (9), because in each term the kth variable is $x_{k+1} \in \mathcal{B}$ due to the variables $x_1, \ldots, x_k$ being shifted by one place to the left. Hence

$$(13) \quad \partial G\chi(x_1, \ldots, x_n) = \sum_{j=k}^{n-1} (-1)^j G\chi(x_1, \ldots, x_j x_{j+1}, \ldots, x_n).$$
$$+ (-1)^n G\chi(x_1, \ldots, x_{n-1})x_n.$$

By (13), $\partial G\chi(x_1,\ldots,x_n)$ is the average over $u \in \mathcal{U}$ of the function $\Psi\colon \mathcal{U} \to \mathcal{V}$ defined by

$$
(14) \qquad \Psi(u) = (-1)^k \chi(x_1,\ldots,x_k x_{k+1} u^*, u, x_{k+2},\ldots,x_n)
$$
$$
+ \sum_{j=k+1}^{n-1} (-1)^j \chi(x_1,\ldots,x_k u^*, u,\ldots,x_j x_{j+1},\ldots,x_n)
$$
$$
+ (-1)^n \chi(x_1,\ldots,x_k u^*, u,\ldots,x_{n-1})x_n.
$$

In the expansion of

$$
\partial\chi(x_1,\ldots,x_{k-1},x_k u^*, u, x_{k+1},\ldots,x_n) = 0,
$$

from the definition of ∂, the first k terms are zero for all $u \in \mathcal{U}$ because each has k^{th} variable in $\mathcal{B}$. Thus

$$
\partial\chi(x_1,\ldots,x_{k-1},x_k u^*, u, x_{k+1},\ldots,x_n) = (-1)^k \chi(x_1,\ldots,x_n)
$$
$$
+ (-1)^{k+1}\chi(x_1,\ldots,x_k u^*, u x_{k+1},\ldots,x_n)
$$
$$
+ \sum_{j=k+1}^{n-1} (-1)^{j+1}\chi(x_1,\ldots,x_k u^*, u,\ldots,x_j x_{j+1},\ldots,x_n)
$$
$$
+ (-1)^n \chi(x_1,\ldots,x_k u^*, u,\ldots,x_{n-1})x_n
$$
$$
= 0,
$$

where in the summation term the original index of summation j has been replaced by $(j+1)$. This substitution makes it clear that this equation may be used to simplify Ψ, using (14), since the summation and final terms are equal up to sign. This yields

$$
(15) \qquad \Psi(u) = (-1)^k \chi(x_1,\ldots,x_k x_{k+1} u^*, u, x_{k+2},\ldots,x_n)
$$
$$
+ (-1)^k \chi(x_1,\ldots,x_n)
$$
$$
+ (-1)^{k+1}\chi(x_1,\ldots,x_k u^*, u x_{k+1},\ldots,x_n).
$$

Since $x_{k+1} \in \mathcal{U}$, the final term in this equation may be rewritten as

$$
(-1)^{k+1}\chi(x_1,\ldots,x_k x_{k+1}(u x_{k+1})^*, u x_{k+1},\ldots,x_n).
$$

Averaging (15) over $u \in \mathcal{U}$ and using the right invariance of μ gives

$$
\int \Psi(u)\,d\mu(u) = (-1)^k \chi(x_1,\ldots,x_n)
$$
$$
= \partial G\chi(x_1,\ldots,x_n).
$$

By (5) and (11) this implies that

$$(\phi - \partial J_{k+1}\phi)(x_1, \ldots, x_n) = 0$$

as required.

Now we estimate $\|J_{k+1}\|$. Observe that $\|G\| \leq 1$ with G mapping from $\mathcal{L}^n(\mathcal{A}, \mathcal{V})$ into $\mathcal{L}^{n-1}(\mathcal{A}, \mathcal{V})$ or from $\mathcal{L}^n_{cb}(\mathcal{A}, \mathcal{V})$ into $\mathcal{L}^{n-1}_{cb}(\mathcal{A}, \mathcal{V})$, because it is an average of maps

$$(x_1, \ldots, x_{n-1}) \to \psi(x_1, \ldots, x_k u^*, u, \ldots, x_n)$$

each with norm no greater than one. Thus the definition (10) of J_{k+1} implies that
$$\|J_{k+1}\| \leq 1 + (n+2)\|J_k\|$$
$$\leq ((n+2)^{k+1} - 1)/(n+1)$$

in both the continuous and completely bounded cases by (4).

The averages are norm limits of partial sums, the unitary group $\mathcal{U}$ is compact, and all the functions are continuous, so both J_1 and G defined in (1) and (6) preserve the subscripts "w", "cb" and "wcb" on $\mathcal{L}$. The map J_{k+1} is inductively defined in terms of G, ∂ and the previous J_k, so also preserves these subscripts. This completes the proof of (a).

Part (b) follows exactly the same line of reasoning, using the weak*-invariant mean average provided by Lemma 3.2.3, except that the subscripts "w" and "wcb" are not preserved. This completes the proof of Lemma 3.2.4.

3.2.5 Remark. The main results in subsequent sections are such that we are really only interested in the amenable group version of the following lemma. There is a version of Lemma 3.2.6 assuming $\mathcal{B}$ to be a finite dimensional C^*-subalgebra and $\mathcal{U}$ to be a compact group, as in Lemma 3.2.4 (a), but this is omitted.

3.2.6 Lemma. *Let $\mathcal{B}$ be the C^*-subalgebra generated by an amenable group $\mathcal{U}$ of unitaries in a unital C^*-subalgebra $\mathcal{A}$ of $B(H)$, and let $\mathcal{V}$ be a dual normal $\mathcal{A}$-module. There is a continuous linear projection Q_n from $\mathcal{L}^n(\mathcal{A}, \mathcal{V})$ onto $\mathcal{L}^n(\mathcal{A}, \mathcal{V}:/\mathcal{B})$ such that $\partial Q_{n-1} = Q_n\partial$. Further the subscript "cb" may be adjoined to $\mathcal{L}$; in both cases $\|Q_n\| \leq 1$.*

Proof. Let $\int \ldots d\mu(u)$ denote the invariant mean as defined on multilinear maps in Lemma 3.2.3. Define the maps $G_j \colon \mathcal{L}^n(\mathcal{A}, \mathcal{V}) \to \mathcal{L}^n(\mathcal{A}, \mathcal{V})$ by the following equations using Lemma 3.2.3 (3), (4), (5) for $j = 0, 1, \ldots, n$.

(1) $\quad (G_0\phi)(x_1, \ldots, x_n) = \displaystyle\int u^*\phi(ux_1, x_2, \ldots, x_n)d\mu(u),$

(2) $\quad (G_k\phi)(x_1, \ldots, x_n) = \displaystyle\int \phi(x_1, \ldots, x_{k-1}, x_k u^*, ux_{k+1}, \ldots, x_n)d\mu(u)$

$(1 \leq k \leq n - 1)$ and

(3) $(G_n\phi)(x_1,\ldots,x_n) = \int \phi(x_1,\ldots,x_n u^*)ud\mu(u)$

for all $\phi \in \mathcal{L}^n(\mathcal{A}, \mathcal{V})$ and $x_1,\ldots,x_n \in \mathcal{A}$.

By Lemma 3.2.3, each G_j is continuous with norm no greater than 1, and has for its image maps that have a $\mathcal{B}$-module property from the jth to the $(j+1)^{\text{th}}$ variable (suitably modified when $j = 0$ and n). Let

$$Q_n = G_n \circ G_{n-1} \circ \cdots \circ G_0.$$

An elementary induction on j shows that $Q_n\phi$ has the $\mathcal{B}$-module property from the jth to $(j+1)^{\text{th}}$ variable for $j = 0,\ldots,n$. Hence $Q_n\phi$ is a $\mathcal{B}$-module map. Further, if ϕ is already a $\mathcal{B}$-module map then $Q_n\phi = \phi$.

The equation $Q_n\partial = \partial Q_{n-1}$ follows from proving by induction that

(4) $$G_k \circ G_{k-1} \ldots G_0 \partial \phi(x_1,\ldots,x_n)$$
$$= x_1 G_{k-1} \ldots G_0 \phi(x_2,\ldots,x_n)$$
$$+ \sum_{1}^{k-1}(-1)^j G_{k-1} \ldots G_0 \phi(x_1,\ldots,x_j x_{j+1},\ldots x_n)$$
$$+ \sum_{k}^{n-1}(-1)^j G_k \ldots G_0 \phi(x_1,\ldots,x_j x_{j+1},\ldots,x_n)$$
$$+ (-1)^n G_k \ldots G_0 \phi(x_1,\ldots,x_{n-1})x_n$$

for $0 \leq k \leq n$, where $x_1,\ldots,x_n \in \mathcal{A}$ and $\phi \in \mathcal{L}^{n-1}(\mathcal{A}, \mathcal{V})$ are fixed. Note G_{-1} is taken to be the identity map on $\mathcal{L}^{n-1}(\mathcal{A}, \mathcal{V})$. The case $k = 0$ is similar to $1 \leq k < n$, so is omitted. We assume (4) for $0 \leq k \leq n - 2$. The crucial idea is that in the first k terms in the sum defining ∂ the variables k and $k + 1$ are moved into the $k - 1$ and k places across which the action of $\mathcal{B}$ commutes. Letting G_{k+1} act on (4) gives the following equation because the first $k + 1$ terms have this left shift on the variables:

(5) $$G_{k+1} \cdot G_k \ldots G_0 \phi(x_1,\ldots,x_n)$$
$$= x_1 G_k \ldots G_0 \phi(x_2,\ldots,x_n)$$
$$+ \sum_{1}^{k-1}(-1)^j G_k \ldots G_0 \phi(x_1,\ldots,x_j x_{j+1},\ldots,x_n)$$
$$+ (-1)^k \int G_k \ldots G_0 \phi(x_1,\ldots,x_k x_{k+1} u^*, ux_{k+2},\ldots,x_n)d\mu(u)$$
$$+ \sum_{k+1}^{n-1}(-1)^j G_{k+1} \ldots G_0 \phi(x_0,\ldots,x_j x_{j+1},\ldots,x_n)$$
$$+ (-1)^n G_{k+1} \ldots G_0 \phi(x_0,\ldots,x_{n-1})x_n.$$

This equals (4) with k replaced by $k+1$ there because the term

$$(-1)^k \int G_k \ldots G_0 \phi(x_1, \ldots, x_k x_{k+1} u^*, u x_{k+2}, \ldots, x_n) d\mu(u)$$

$$= (-1)^k G_k \ldots G_0 \phi(x_1, \ldots, x_k x_{k+1}, x_{k+2}, \ldots, x_n)$$

because the action of G_k ensures that the two "integrands" are equal. The equation $Q_n \partial = \partial Q_{n-1}$ now follows from (4) with $k = n$ by a further application of G_n using $G_n \circ G_n = G_n$. This proves Lemma 3.2.6.

3.2.7 Theorem. *Let $\mathcal{B}$ be the C^*-subalgebra generated by an amenable group $\mathcal{U}$ of unitaries in a unital C^*-subalgebra $\mathcal{A}$ of $B(H)$, and let $\mathcal{V}$ be a dual normal $\mathcal{A}$-module. Then*

$$H^n(\mathcal{A}, \mathcal{V}) \cong H^n(\mathcal{A}, \mathcal{V}: /\mathcal{B}),$$

$$H_{cb}^n(\mathcal{A}, \mathcal{V}) \cong H_{cb}^n(\mathcal{A}, \mathcal{V}: /\mathcal{B}),$$

for all $n \in \mathbb{N}$ with the isomorphism induced by the natural embedding of $\mathcal{L}^n(\mathcal{A}, \mathcal{V}: /\mathcal{B})$ into $\mathcal{L}^n(\mathcal{A}, \mathcal{V})$.

Proof. The natural embedding clearly induces a homomorphism from $H^n(\mathcal{A}, \mathcal{V}: /\mathcal{B})$ into $H^n(\mathcal{A}, \mathcal{V})$; it remains to show that it is an isomorphism. The map is surjective by Lemma 3.2.4 (a), which ensures that $\phi - \partial K_n \phi$ is a $\mathcal{B}$-module map. It is also injective by Lemma 3.2.6, since if $\phi \in \mathcal{L}^n(\mathcal{A}, \mathcal{V}: /\mathcal{B})$ and $\phi = \partial \psi$, then

$$\phi = Q_n \phi = Q_n \partial \psi = \partial Q_{n-1} \psi,$$

and $Q_{n-1} \psi \in \mathcal{L}^{n-1}(\mathcal{A}, \mathcal{V}: /\mathcal{B})$. Exactly the same argument works in the completely bounded situation.

3.3 Norm Continuous and Normal Cohomology are Equal: Extension Techniques

The main result of this section is the following theorem, which states that normal cohomology and norm continuous cohomology are equal for dual normal modules. Throughout this section we shall consider a C^*-algebra $\mathcal{A}$ as a subalgebra of $B(H)$, so $\mathcal{A}$ has a natural weak and ultraweak topology inherited from $B(H)$.

3.3.1 Theorem. *Let $\mathcal{A}$ be a C^*-subalgebra of $B(H)$ with weak closure $\bar{\mathcal{A}}$.*

(1) If $\mathcal{V}$ is a dual normal $\bar{\mathcal{A}}$-module, then

$$H_c^n(\mathcal{A}, \mathcal{V}) \cong H_w^n(\mathcal{A}, \mathcal{V}) \cong H_w^n(\bar{\mathcal{A}}, \mathcal{V})$$

for all $n \in \mathbb{N}$.

(2) If $\mathcal{V}$ is a subspace of $B(H)$ and is a normal $\bar{\mathcal{A}}$-module with the product from $B(H)$, then

$$H_{cb}^n(\mathcal{A}, \mathcal{V}) \cong H_{wcb}^n(\mathcal{A}, \mathcal{V}) \cong H_{wcb}^n(\bar{\mathcal{A}}, \mathcal{V})$$

for all $n \in \mathbb{N}$.

Further, if $\mathcal{B}$ is a C^*-subalgebra of $\mathcal{A}$, then these isomorphisms all hold with $\mathcal{B}$-module maps throughout.

The proof of this theorem will be given in (3.3.6) based on an extension lemma (3.3.2), a lemma that ensures that a continuous cocycle can be modified by a continuous coboundary so as to be normal (3.3.5), and on a lemma (3.3.4) showing that a normal coboundary arising from a continuous map already arises from a normal map.

3.3.2 Lemma. *Let $\mathcal{A}$ and $\mathcal{B}$ be C^*-algebras acting on a Hilbert space H with ultraweak closures $\bar{\mathcal{A}}$ and $\bar{\mathcal{B}}$ (respectively), and let τ be a bounded bilinear form on $\mathcal{A} \times \mathcal{B}$. If τ is separately ultraweakly continuous, then τ extends uniquely to a separately ultraweakly continuous bilinear form $\bar{\tau}$ on $\bar{\mathcal{A}} \times \bar{\mathcal{B}}$.*

Proof. By the Kaplansky density theorem, for each fixed b in $\mathcal{B}$, the ultraweakly continuous linear functional $a \mapsto \tau(a, b)$ on $\mathcal{A}$ extends without change of norm to an ultraweakly continuous linear functional $T(b)$ on $\bar{\mathcal{A}}$. The mapping $b \mapsto T(b)$ is a bounded linear map from $\mathcal{B}$ into the predual $(\bar{\mathcal{A}})_*$ of $\bar{\mathcal{A}}$ with $\|T\| \leq \|\tau\|$. By its definition, $\langle a, Tb \rangle = \tau(a, b)$, where $\langle \cdot, \cdot \rangle$ denotes the pairing between $\bar{\mathcal{A}}$ and its predual $(\bar{\mathcal{A}})_*$. Since τ is ultraweakly continuous in its second argument with the first fixed, T is a continuous linear map from $\mathcal{B}$ with the ultraweak topology into $(\bar{\mathcal{A}})_*$ with the weak topology $\sigma((\bar{\mathcal{A}})_*, \mathcal{A})$ induced by elements of $\mathcal{A}$. We would like to replace this topology with the finer $\sigma((\bar{\mathcal{A}})_*, \bar{\mathcal{A}})$ topology, but this requires a further argument which we now give. By [T, Theorem 5.4]; [Ak, Cor. II.9], $T(\mathcal{B}_1)$ is a relatively compact subset of $(\bar{\mathcal{A}})_*$ in the $\sigma((\bar{\mathcal{A}})_*, \bar{\mathcal{A}})$ topology, where $\mathcal{B}_1$ denotes the closed unit ball of $\mathcal{B}$. Hence the $\sigma((\bar{\mathcal{A}})_*, \bar{\mathcal{A}})$ topology coincides with the coarser $\sigma((\bar{\mathcal{A}})_*, \mathcal{A})$ topology. This and the continuity of T above imply that T is continuous as a map from $\mathcal{B}_1$ with the ultraweak topology into $(\bar{\mathcal{A}})_*$ with the $\sigma((\bar{\mathcal{A}})_*, \bar{\mathcal{A}})$ topology. So for each fixed $a \in \bar{\mathcal{A}}$, the linear functional $b \to \langle a, Tb \rangle$ on $\mathcal{B}$ is ultraweakly continuous on the unit ball $\mathcal{B}_1$, and hence on $\mathcal{B}$. Define the form $\bar{\tau}$ on $\bar{\mathcal{A}} \times \mathcal{B}$ by

$$\bar{\tau}(a, b) = \langle a, Tb \rangle$$

for all $a \in \bar{\mathcal{A}}$ and $b \in \mathcal{B}$; this has the required properties.

3.3.3 Lemma. *Let $\mathcal{A}$ be a C^*-algebra acting on a Hilbert space H and let $\mathcal{V}$ be the dual of a Banach space $\mathcal{V}_*$. If ϕ is a bounded n-linear map from $\mathcal{A} \times \cdots \times \mathcal{A}$ into $\mathcal{V}$ that is separately continuous relative to the ultraweak topology on $\mathcal{A}$ and the weak* topology on $\mathcal{V}$, then ϕ extends uniquely, without change of norm, to a bounded n-linear map $\bar{\phi}$ from $\bar{\mathcal{A}} \times \cdots \times \bar{\mathcal{A}}$ to $\mathcal{V}$, which is continuous relative to the ultraweak topology on $\bar{\mathcal{A}}$ and the weak* topology on $\mathcal{V}$. If, in addition, $\mathcal{V}$ is a subspace of $B(H)$ and ϕ is completely bounded, then $\bar{\phi}$ is completely bounded and $\|\bar{\phi}\|_{cb} = \|\phi\|_{cb}$.*

Proof. Firstly this will be proved for bounded n-linear forms, that is, when $\mathcal{V}$ is the complex numbers $\mathbf{C}$, and the general and completely bounded cases will be deduced from this. Let $\mathcal{V} = \mathbf{C}$ for the present.

A finite sequence $\phi_0 = \phi, \phi_1, \ldots, \phi_n$ of n-linear functionals is constructed with the property that ϕ_k is defined on $\bar{\mathcal{A}} \times \cdots \times \bar{\mathcal{A}} \times \mathcal{A} \times \cdots \times \mathcal{A}$ with k copies of $\bar{\mathcal{A}}$ and $n - k$ copies of $\mathcal{A}$, extends ϕ_{k-1} without change of norm when $k \geq 1$ and is separately ultraweakly continuous. This will prove the existence of $\bar{\phi} = \phi_n$; the uniqueness of $\bar{\phi}$ follows directly from the ultraweak continuity and the ultraweak density of $\mathcal{A}$ in $\bar{\mathcal{A}}$.

Suppose that $\phi_0, \ldots, \phi_{k-1}$ have been constructed ($1 \leq k \leq n$). Keeping all the a_j's fixed in their domains for ϕ_{k-1} for $j \neq k$, and varying a_k gives a linear functional

$$a_k \to \phi_{k-1}(a_1, \ldots, a_k, \ldots, a_n)$$

which is ultraweakly continuous on $\mathcal{A}$. The norm of this functional does not exceed

$$\|\phi_{k-1}\| \, \|a_1\| \cdots \|a_{k-1}\| \, \|a_{k+1}\| \cdots \|a_n\|.$$

By the Kaplansky density theorem, this functional extends without change of norm to an ultraweakly continuous linear functional on $\bar{\mathcal{A}}_k$, which we denote by

$$a_k \to \phi_k(a_1, \ldots, a_n).$$

Clearly ϕ_k is an n-linear form on $\bar{\mathcal{A}} \times \cdots \times \bar{\mathcal{A}} \times \mathcal{A} \times \cdots \times \mathcal{A}$ with k copies of $\bar{\mathcal{A}}$ and $n - k$ copies of $\mathcal{A}$, that extends ϕ_{k-1} without change of norm, and is ultraweakly continuous in its kth argument. We need to prove the ultraweak continuity of ϕ_k in its other arguments for $a_k \in \bar{\mathcal{A}} \backslash \mathcal{A}$.

Let $1 \leq j \leq n$ with $j \neq k$, and fix a_i for all $i \neq j, k$ with a_i in $\bar{\mathcal{A}}$ for $i < k$ or in $\mathcal{A}$ for $i > k$. Let $\mathcal{B} = \bar{\mathcal{A}}$ if $j < k$ and $\mathcal{B} = \mathcal{A}$ if $j > k$. Let τ be the bounded bilinear form defined on $\mathcal{A} \times \mathcal{B}$ by

$$\tau(a_k, a_j) = \phi_{k-1}(a_1, \ldots, a_n) = \phi_k(a_1, \ldots, a_n).$$

This form is separately ultraweakly continuous by the assumptions concerning ϕ_{k-1}, and hence extends to a bounded bilinear form $\bar{\tau}$ on $\bar{\mathcal{A}} \times \mathcal{B}$ which is separately ultraweakly continuous by Lemma 3.3.2. By the ultraweak continuity of both $\bar{\tau}(a_k, a_j)$ and $\phi_k(a_1, \ldots, a_n)$ in the variable $a_k \in \bar{\mathcal{A}}$ and their equality on $\mathcal{A}$, it follows that

$$\bar{\tau}(a_k, a_j) = \phi_k(a_1, \ldots, a_n)$$

on $\bar{\mathcal{A}} \times \mathcal{B}$. This shows that ϕ_k is ultraweakly continuous in $a_j \in \mathcal{B}$ for each $a_k \in \bar{\mathcal{A}}$, because $\bar{\tau}$ has this property. This finishes the inductive proof for $\mathcal{V} = \mathbb{C}$.

Let $\mathcal{V}$ be the dual of a Banach space $\mathcal{V}_*$. For each $w \in \mathcal{V}_*$, the n-linear form $\rho(w)$ defined on $\mathcal{A}^n$ by

$$\rho(w)(a_1, \ldots, a_n) = \langle \phi(a_1, \ldots, a_n), w \rangle$$

is bounded by $\|\varphi\| \cdot \|w\|$ and is separately ultraweakly continuous. By what has just been proved above, $\rho(\omega)$ extends uniquely, without change of norm, to a separately ultraweakly continuous n-linear form $\overline{\rho(w)}$ on $(\bar{\mathcal{A}})^n$. This shows that the map $w \mapsto \overline{\rho(w)}(a_1, \ldots, a_n)$ is a bounded linear functional on $\mathcal{V}_*$ for all $a_j \in \bar{\mathcal{A}}$, so defines an element in $\mathcal{V} = (\mathcal{V}_*)^*$ denoted by $\bar{\phi}(a_1, \ldots, a_n)$. The multilinearity of all the constructions yields that $\bar{\phi}$ is an n-linear map from $(\bar{\mathcal{A}})^n$ into $\mathcal{V}$, which is separately ultraweakly-weak* continuous and has $\|\bar{\phi}\| = \|\phi\|$.

One detail remains: if ϕ is completely bounded we need to show that $\bar{\phi}$ is completely bounded. Let k be a positive integer. The ultraweak density of the unit ball of $M_k(\mathcal{A})$ in the unit ball of $M_k(\bar{\mathcal{A}})$, the Kaplansky density theorem and the definition of the k-fold amplification $(\bar{\phi})_k$ as a matrix product imply that $(\bar{\phi})_k(a_1, \ldots, a_n)$, for $a_j \in M_k(\bar{\mathcal{A}})$, is a weak limit of elements $\phi_k(b_1, \ldots, b_n)$, where $b_j \in M_k(\mathcal{A})$ with $\|b_j\| \leq \|a_j\|$. This implies that

$$\|(\bar{\phi})_k\| \leq \|\phi_k\| \leq \|\phi\|_{cb},$$

from which $\|\bar{\phi}\|_{cb} = \|\phi\|_{cb}$ follows since $\bar{\phi}$ is an extension of ϕ. This proves Lemma 3.3.3.

The crucial link between continuous and normal cohomology depends on passing to a subalgebra of the enveloping von Neumann algebra and cutting down by a suitable central projection. This step is separated out in the definition and properties of three maps S_n, T_n, and W_n in the following lemma. The technical parts of Lemma 3.3.4 are used in the proof of Lemma 3.3.5, Theorem 3.3.1 and in Section 3.4. The two Lemmas 3.2.4 and 3.3.4 form the technical foundation of this chapter.

Recall that the weak closure $\pi(\mathcal{A})^-$ of the universal representation π of $\mathcal{A}$ is the enveloping von Neumann algebra $\mathcal{A}^{**}$ of $\mathcal{A}$ [Di2, T].

3.3.4 Lemma. *Let $\mathcal{A}$ be a C^*-algebra acting on a Hilbert space H with weak closure $\bar{\mathcal{A}}$ and let $\mathcal{V}$ be a dual normal $\bar{\mathcal{A}}$-module. If π is the universal representation of $\mathcal{A}$, then $\mathcal{V}$ may be regarded as a dual normal $\pi(\mathcal{A})^-$-module in a natural way using the minimal central projection p in $\pi(\mathcal{A})^-$ such that*

$$\pi(\mathcal{A})^- \cdot p = \bar{\mathcal{A}}.$$

There are continuous linear maps

$$T_n \colon \mathcal{L}^n(\mathcal{A}, \mathcal{V}) \to \mathcal{L}^n_w(\pi(\mathcal{A})^-, \mathcal{V}),$$
$$S_n \colon \mathcal{L}^n_w(\pi(\mathcal{A})^-, \mathcal{V}) \to \mathcal{L}^n_w(\bar{\mathcal{A}}, \mathcal{V}),$$
$$W_n \colon \mathcal{L}^n_w(\pi(\mathcal{A})^-, \mathcal{V}) \to \mathcal{L}^n(\mathcal{A}, \mathcal{V}),$$

such that

(1) $\partial T_n = T_{n+1}\partial$ and $\partial S_n = S_{n+1}\partial$,

(2) each preserves complete boundedness as follows:
$$T_n \colon \mathcal{L}^n_{cb}(\mathcal{A}, \mathcal{V}) \to \mathcal{L}^n_{wcb}(\pi(\mathcal{A})^-, \mathcal{V}),$$

$$S_n \colon \mathcal{L}^n_{wcb}(\pi(\mathcal{A})^-, \mathcal{V}) \to \mathcal{L}^n_{wcb}(\bar{\mathcal{A}}, \mathcal{V}),$$

$$W_n \colon \mathcal{L}^n_{wcb}(\pi(\mathcal{A})^-, \mathcal{V}) \to \mathcal{L}^n_{cb}(\mathcal{A}, \mathcal{V}),$$

and in both the continuous and completely bounded cases $\|T_n\|$, $\|S_n\|$, $\|W_n\| \le 1$,

(3) if $\mathcal{B}$ is a C^*-subalgebra of $\mathcal{A}$, T_n maps $\mathcal{B}$-module maps to $\pi(\mathcal{B})^-$-module maps, and S_n and W_n map $\pi(\mathcal{B})^-$-module maps to $\mathcal{B}$-module maps,

(4) $S_n T_n$ is a projection from $\mathcal{L}^n(\mathcal{A}, \mathcal{V})$ onto $\mathcal{L}^n_w(\mathcal{A}, \mathcal{V})$,

(5) if $\mathcal{C}$ is the C^*-algebra generated by 1 and p, and if

$$\psi \in \mathcal{L}^n_w(\pi(\mathcal{A})^-, \mathcal{V}: /\mathcal{C}),$$

then $W_n \psi = S_n \psi \in \mathcal{L}^n_w(\bar{\mathcal{A}}, \mathcal{V})$,

(6) $W_n T_n$ is the identity map on $\mathcal{L}^n(\mathcal{A}, \mathcal{V})$.

Proof. The proof consists of turning $\mathcal{V}$ into a dual normal $\pi(\mathcal{A})^-$-module then giving the definitions and properties of T_n, S_n and W_n. The interplay between these three maps is important subsequently.

Since π is the universal representation of $\mathcal{A}$, there is a projection p in the centre of the weak closure $\pi(\mathcal{A})^-$ of $\pi(\mathcal{A})$ and an isomorphism θ from $p\pi(\mathcal{A})^-$ onto $\bar{\mathcal{A}}$ such that

(1) $$\theta(p\pi(a)) = a \quad \text{and} \quad \theta(px) = \pi^{-1}(x)$$

for all $a \in \mathcal{A}$ and $x \in \pi(\mathcal{A})$ [Di2,T]. Now θ is a homeomorphism from $p\pi(\mathcal{A})^-$ onto $\bar{\mathcal{A}}$ if both have their ultraweak topologies, since *-isomorphisms between von Neumann algebras are ultraweak homeomorphisms [Di2,T]. Now define $x \cdot v$ and $v \cdot x$ by

$$(2) \qquad x \cdot v = \theta(px)v \quad \text{and} \quad v \cdot x = v\theta(px)$$

for all $x \in \pi(\mathcal{A})^-$ and $v \in \mathcal{V}$. With these definitions $\mathcal{V}$ becomes a dual normal $\pi(A)^-$-module, since it is a dual normal $\bar{A}$-module. Observe that

$$(3) \qquad p \cdot v = v \cdot p = v \quad \text{for all} \quad v \in \mathcal{V}.$$

The equation

$$(4) \qquad \phi_1(x_1, \ldots, x_n) = \phi(\theta(px_1), \ldots, \theta(px_n))$$

for all $x_1, \ldots, x_n \in \pi(\mathcal{A})$ defines $\phi_1 \in \mathcal{L}^n(\pi(\mathcal{A}), \mathcal{V})$. This map is separately ultraweakly-weak* continuous in each of its arguments, because π is the universal representation of $\mathcal{A}$, so each continuous linear functional on $\pi(\mathcal{A})$ is ultraweakly continuous [T, Theorem 2.4]; that is $\phi_1 \in \mathcal{L}_w^n(\pi(\mathcal{A}), \mathcal{V})$. Lemma 3.3.3 extends ϕ_1 to $\bar{\phi}_1$ in $\mathcal{L}_w^n(\pi(\mathcal{A})^-, \mathcal{V})$ without change of norm. Define $T_n \colon \mathcal{L}^n(\mathcal{A}, \mathcal{V}) \to \mathcal{L}_w^n(\pi(\mathcal{A})^-, \mathcal{V})$ by $T_n\phi = \bar{\phi}_1$. The map T_n is easily seen to be an isometry in the continuous and completely bounded cases.

Let $\phi \in \mathcal{L}^n(\mathcal{A}, \mathcal{V})$. To show that

$$\partial T_n \phi = T_{n+1} \partial \phi$$

it is sufficient to prove that

$$\partial T_n \phi(x_1, \ldots, x_{n+1}) = T_{n+1}\partial\phi(x_1, \ldots, x_{n+1})$$

for all $x_1, \ldots, x_{n+1} \in \pi(\mathcal{A})$ by the separate ultraweak-weak* continuity of $\partial T_n \phi$ and $T_{n+1}\partial\phi$. The definition of T_n, combined with (4) and (2), implies that

$$\begin{aligned}
\partial T_n \phi(x_1, \ldots, x_{n+1}) =\,& \theta(px_1)\phi(\theta(px_2), \ldots, \theta(px_{n+1})) \\
& + \sum_{j=1}^{n}(-1)^j \phi(\ldots, \theta(px_j)\theta(px_{j+1}), \ldots) \\
& + (-1)^{n+1}\phi(\theta(px_1), \ldots, \theta(px_n))\theta(px_{n+1}) \\
=\,& T_{n+1}\partial\phi(x_1, \ldots, x_{n+1})
\end{aligned}$$

for all $x_1, \ldots, x_{n+1} \in \pi(\mathcal{A})$, because

$$\theta(px_j)\theta(px_{j+1}) = \theta(px_j px_{j+1}) = \theta(px_j x_{j+1}),$$

which holds since p is a central projection in $\pi(\mathcal{A})^-$.

Finally, note that if $\phi \in \mathcal{L}^n(\mathcal{A}, \mathcal{V}: /B)$, then $T_n \phi \in \mathcal{L}_w^n(\pi(\mathcal{A})^-, \mathcal{V}: /\pi(B)^-)$ because of the definition (3) of the module action, and p being in the centre of $\pi(\mathcal{A})^-$. For example, by the ultraweak-weak* continuity of the maps involved, the following calculation is sufficient to prove the module action across the j, $(j+1)$ positions. If $x_1, \ldots, x_n \in \pi(\mathcal{A})$ and $b \in B$ with $x_j = \pi(a_j)$, then

$$
\begin{aligned}
&T_n\phi(x_1, \ldots, x_j\pi(b), x_{j+1}, \ldots, x_n) \\
&= \phi(\theta(px_1), \ldots, \theta(px_j\pi(b)), \ldots, \theta(px_n)) \\
&= \phi(\theta(px_1), \ldots, a_j b, a_{j+1}, \ldots, \theta(px_n)) \\
&= \phi(\theta(px_1), \ldots, a_j, ba_{j+1}, \ldots, \theta(px_n)) \\
&= \phi(\theta(px_1), \ldots, \theta(px_j), \theta(p\pi(b)x_{j+1}), \ldots, \theta(px_n)) \\
&= T_n\phi(x_1, \ldots, x_j, \pi(b)x_{j+1}, \ldots, x_n)
\end{aligned}
$$

as required. The properties of T_n have now been checked.

The map $S_n\colon \mathcal{L}_w^n(\pi(\mathcal{A})^-, \mathcal{V}) \to \mathcal{L}_w^n(\bar{\mathcal{A}}, \mathcal{V})$ is defined by

$$
(5) \qquad (S_n\psi)(a_1, \ldots, a_n) = \psi(\theta^{-1}(a_1), \ldots, \theta^{-1}(a_n))
$$

for all $a_1, \ldots, a_n \in \bar{\mathcal{A}}$. Note that since ψ and θ^{-1} are normal by hypothesis and (1), $S_n\psi$ is normal. Further $\|S_n\| \leq 1$ in the continuous and completely bounded cases. Now $\theta(p\theta^{-1}(a)) = a$ by (1) and $p = p^2$, and

$$
\begin{aligned}
\theta^{-1}(a) \cdot v &= \theta(p\theta^{-1}(a))v = av, \\
v \cdot \theta^{-1}(a) &= va
\end{aligned}
$$

for all $a \in \mathcal{A}$ and $v \in \mathcal{V}$. Hence

$$
\begin{aligned}
S_{n+1}\partial\psi(a_1, \ldots, a_{n+1}) &= \partial\psi(\theta^{-1}(a_1), \ldots, \theta^{-1}(a_{n+1})) \\
&= a_1\psi(\theta^{-1}(a_2), \ldots, \theta^{-1}(a_{n+1})) \\
&\quad + \sum_1^n (-1)^j \psi(\theta^{-1}(a_1), \ldots, \theta^{-1}(a_j a_{j+1}), \ldots, \theta^{-1}(a_{n+1})) \\
&\quad + (-1)^{n+1}\psi(\theta^{-1}(a_1), \ldots, \theta^{-1}(a_n))a_{n+1} \\
&= \partial S_n\psi(a_1, \ldots, a_{n+1})
\end{aligned}
$$

for all $a_1, \ldots, a_{n+1} \in \mathcal{A}$. The normality of the maps ensures that equality holds on $\bar{\mathcal{A}}$. If ψ is a $\pi(B)^-$-module map, then clearly $S_n\psi$ is a $\bar{B}$-module map.

The map $W_n\colon \mathcal{L}^n(\pi(\mathcal{A})^-, \mathcal{V}) \to \mathcal{L}^n(\mathcal{A}, \mathcal{V})$ is defined by

$$
W_n\psi(a_1, \ldots, a_n) = \psi(\pi(a_1), \ldots, \pi(a_n))
$$

for all $a_1, \ldots, a_n \in \mathcal{A}$. Clearly W_n is a continuous linear map with $\|W_n\| \leq 1$ in both the continuous and completely bounded cases. If $\phi \in \mathcal{L}^n(\mathcal{A}, \mathcal{V})$, then

$$
\begin{aligned}
W_n T_n \phi(a_1, \ldots, a_n) &= T_n \phi(\pi(a_1), \ldots, \pi(a_n)) \\
&= \phi(\theta(p\pi(a_1)), \ldots, \theta(p\pi(a_n))) \\
&= \phi(a_1, \ldots, a_n)
\end{aligned}
$$

by the definitions of W_n, T_n, and θ, (1) and (4). This proves (6).

Let ψ be in $\mathcal{L}^n_w(\pi(\mathcal{A})^-, \mathcal{V}: /\mathcal{C})$. Then

$$
\begin{aligned}
(W_n \psi)(a_1, \ldots, a_n) &= \psi(\pi(a_1), \ldots, \pi(a_n)) \text{ by definition of } W_n \\
&= \psi(\pi(a_1), \ldots, \pi(a_n)) \cdot p \text{ by (3)} \\
&= \psi(\pi(a_1)p, \ldots, \pi(a_n)p)
\end{aligned}
$$

because ψ is a $\mathcal{C}$-module map and $p = p^2$ is in the centre of $\pi(\mathcal{A})^-$. Since $\theta^{-1}(a) = \pi(a)p$ for all $a \in \mathcal{A}$ by (1),

$$
\begin{aligned}
(W_n \psi)(a_1, \ldots, a_n) &= \psi(\theta^{-1}(a_1), \ldots, \theta^{-1}(a_n)) \\
&= S_n \psi(a_1, \ldots, a_n).
\end{aligned}
$$

This proves (5).

3.3.5 Lemma. *Let $\mathcal{A}$ be a C^*-algebra acting on a Hilbert space H with weak closure $\bar{\mathcal{A}}$ and let $\mathcal{V}$ be a dual normal $\bar{\mathcal{A}}$-module. There is a continuous linear map $J_n: \mathcal{L}^n(\mathcal{A}, \mathcal{V}) \to \mathcal{L}^{n-1}(\mathcal{A}, \mathcal{V})$ with*

$$
\|J_n\| \leq ((n+2)^n - 1)/(n+1),
$$

such that if $\phi \in \mathcal{L}^n(\mathcal{A}, \mathcal{V})$ with $\partial \phi = 0$, then $\phi - \partial J_n \phi \in \mathcal{L}^n_w(\mathcal{A}, \mathcal{V})$. The map J_n has the following two properties:

(1) J_n maps $\mathcal{L}^n_{cb}(\mathcal{A}, \mathcal{V})$ into $\mathcal{L}^{n-1}_{cb}(\mathcal{A}, \mathcal{V})$ and has the same bound on the norm between these spaces,

(2) if $\mathcal{B}$ is a C^*-subalgebra of $\mathcal{A}$, then

$$
J_n: \mathcal{L}^n(\mathcal{A}, \mathcal{V}: /\mathcal{B}) \to \mathcal{L}^{n-1}(\mathcal{A}, \mathcal{V}: /\mathcal{B})
$$

and similarly with the subscript "cb" on both $\mathcal{L}$'s.

Proof. The maps T_n and W_n of Lemma 3.3.4 are used together with the map K_n of Lemma 3.2.4 to define J_n. Let π be the universal representation of $\mathcal{A}$ and let p be the central projection in $\pi(\mathcal{A})^-$ such that $p\pi(\mathcal{A})^-$ is naturally isomorphic to $\bar{\mathcal{A}}$ (see (3.3.4)).

The unitary subgroup consisting of the two elements $\{1, 2p - 1\}$ generates a two dimensional C^*-subalgebra $\mathcal{C}$ in the centre of $\pi(\mathcal{A})^-$ to which the averaging techniques of Lemma 3.2.4(a) are applied. This gives a continuous linear map $K_n\colon \mathcal{L}_w^n(\pi(\mathcal{A})^-, \mathcal{V}) \to \mathcal{L}_w^{n-1}(\pi(\mathcal{A})^-, \mathcal{V})$ such that $(I - \partial K_n)\psi$ is a $\mathcal{C}$-module map for each $\psi \in \mathcal{L}_w^n(\pi(\mathcal{A})^-, \mathcal{V})$ with $\partial\psi = 0$. Define $J_n\colon \mathcal{L}^n(\mathcal{A}, \mathcal{V}) \to \mathcal{L}^{n-1}(\mathcal{A}, \mathcal{V})$ by $J_n = W_{n-1}K_nT_n$. Then $\|J_n\| \leq ((n+2)^n - 1)/(n+1)$ in the continuous and completely bounded norms and J_n takes $\mathcal{B}$-module maps to $\mathcal{B}$-module maps by Lemmas 3.2.4 and 3.3.4. Since $\partial W_{n-1} = W_n\partial$ and W_nT_n is the identity on $\mathcal{L}^n(\mathcal{A}, \mathcal{V})$ (Lemma 3.3.4),

$$(1) \qquad \phi - \partial J_n\phi = \phi - \partial W_{n-1}K_nT_n\phi = W_n(T_n\phi - \partial K_nT_n\phi).$$

Since

$$\partial T_n\phi = T_{n+1}\partial\phi = 0,$$

it follows that $T_n\phi - \partial K_nT_n\phi$ is a $\mathcal{C}$-module map. Now W_n takes $\mathcal{C}$-module maps to normal maps (3.3.4 (5)) so $\phi - \partial J_n\phi$ is normal.

3.3.6 Proof of Theorem 3.3.1. Lemma 3.3.3 shows that the restriction map from $\mathcal{L}_w^n(\bar{\mathcal{A}}, \mathcal{V})$ to $\mathcal{L}_w^n(\mathcal{A}, \mathcal{V})$ is an isomorphism so that

$$H_w^n(\mathcal{A}, \mathcal{V}) \cong H_w^n(\bar{\mathcal{A}}, \mathcal{V}),$$

with similar isomorphisms in the completely bounded and $\mathcal{B}$-module map cases. If $\phi \in \mathcal{L}_w^n(\mathcal{A}, \mathcal{V})$ with $\phi = \partial\psi$ for some $\psi \in \mathcal{L}^{n-1}(\mathcal{A}, \mathcal{V})$, then

$$\phi = S_nT_n\partial\psi = \partial S_nT_n\psi$$

with $S_nT_n\psi \in \mathcal{L}_w^{n-1}(\mathcal{A}, \mathcal{V})$ by Lemma 3.3.4 (1) and (4). Thus the natural embedding of $\mathcal{L}_w^n(\mathcal{A}, \mathcal{V})$ into $\mathcal{L}^n(\mathcal{A}, \mathcal{V})$ induces an injective map on the cohomology groups. Lemma 3.3.5 shows that this map is surjective, so

$$H_w^n(\mathcal{A}, \mathcal{V}) \cong H^n(\mathcal{A}, \mathcal{V}).$$

The results also hold in the completely bounded and $\mathcal{B}$-module map cases by the respective lemmas.

The following result is useful in splitting the cohomology groups of von Neumann algebras into direct sums according to the central direct sum splitting of the algebra itself into various types.

3.3.7 Theorem. *Let* $\mathcal{M} = \mathcal{M}_1 \oplus \mathcal{M}_2$ *be a direct sum splitting of* $\mathcal{M}$ *into von Neumann algebras* $\mathcal{M}_1$ *and* $\mathcal{M}_2$ *with the identity 1 of* $\mathcal{M}$ *decomposing as* $1 = e_1 \oplus e_2$.

(1) If V is a dual normal $\mathcal{M}$-module, then

$$H_w^n(\mathcal{M}, V) \cong H_w^n(\mathcal{M}_1, V) \oplus H_w^n(\mathcal{M}_2, V)$$
$$\cong H_w^n(\mathcal{M}_1, e_1 V e_1) \oplus H_w^n(\mathcal{M}_2, e_2 V e_2).$$

(2) If $\mathcal{M} \subseteq B(H)$, and if V is both a dual normal $\mathcal{M}$-module and a subspace of $B(H)$, then (1) holds with the subscript "wcb" in place of "w".

Proof. Let $\mathcal{B}$ be the $*$-subalgebra of $\mathcal{M}$ generated by e_1 and e_2. Note that $\mathcal{B}$ is contained in the centre of $\mathcal{M}$ and has dimension two. By Theorems 3.2.7 and 3.3.1, $H_w^n(\mathcal{M}, V) \cong H_w^n(\mathcal{M}, V: /\mathcal{B})$. The required isomorphism is now defined at the operator space level by

$$\phi \mapsto \phi|_{\mathcal{M}_1} \oplus \phi|_{\mathcal{M}_2} = e_1 \phi|_{\mathcal{M}_1} \oplus e_2 \phi|_{\mathcal{M}_2}$$

from $\mathcal{L}_w^n(\mathcal{M}, V: /\mathcal{B})$ into $\mathcal{L}_w^n(\mathcal{M}_1, V) \oplus \mathcal{L}_w^n(\mathcal{M}_2, V)$ and into $\mathcal{L}_2^n(\mathcal{M}_1, e_1 V e_1) \oplus \mathcal{L}_w^n(\mathcal{M}_2, e_2 V e_2)$.

Note that for $\phi \in \mathcal{L}_w^n(\mathcal{M}, V: /\mathcal{B})$

$$\phi(x, y) = \sum_{j=1}^{2} \phi(e_j x, y e_j) = \sum_{j=1}^{2} e_j \phi(x, y) e_j$$

for $x, y \in \mathcal{M}_j$ and $j = 1, 2$. The result now follows directly.

Note that the above result also holds for "c" and "cb" replaced by "w" and "wcb".

The following corollary is a consequence of this theorem by the Murray–von Neumann decomposition of a von Neumann algebra into its central direct summands of different types.

3.3.8 Corollary. *Let $\mathcal{M}$ be a von Neumann algebra and let*

$$\mathcal{M} = \mathcal{M}_I \oplus \mathcal{M}_{II_1} \oplus \mathcal{M}_{II_\infty} \oplus \mathcal{M}_{III}$$

be the central direct summand decomposition of $\mathcal{M}$ into von Neumann algebras of types I, II_1, II_∞, III.

(1) If V is a dual normal $\mathcal{M}$-module, then

$$H_w^n(\mathcal{M}, V) \cong H_w^n(\mathcal{M}_{II_1}, V) \oplus H_w^n(\mathcal{M}_{II_\infty}, V) \oplus H_w^n(\mathcal{M}_{III}, V).$$

(2) If V is a dual normal $\mathcal{M}$-module and a subspace of $B(H)$, where H is the Hilbert space on which $\mathcal{M}$ acts, then "w" may be replaced by "wcb" in (1).

Proof. Type I von Neumann algebras are hyperfinite, and we will show in Corollary 3.4.6 that the cohomology groups of such algebras are zero. Thus $H_w^n(\mathcal{M}_I, \mathcal{V}) = 0$, and the result follows from Theorem 3.3.7.

3.4 Averaging in Normal Cohomology

The main result of this section and of the whole chapter (3.4.1), ensures that the normal (completely bounded) cohomology of a von Neumann algebra over a dual normal module is isomorphic to the cohomology that is zero on a hyperfinite von Neumann subalgebra. This is the crucial result used in the calculations in subsequent chapters of these notes. The proof consists of combining the lemmas of Sections 3.2 and 3.3. Averaging a normal cocycle by an amenable subgroup yields a cocycle that is not necessarily normal (Section 3.2). The averaged cocycle is adjusted as in Section 3.3 to be normal and yields the cocycle zero on the hyperfinite subalgebra.

3.4.1 Theorem. *Let $\mathcal{M}$ be a von Neumann algebra with a hyperfinite subalgebra $\mathcal{N}$ and let $\mathcal{M}$ act on H.*
(1) If $\mathcal{V}$ is a dual normal $\mathcal{M}$-module, then

$$H_w^n(\mathcal{M}, \mathcal{V}: /\mathcal{N}) = H_w^n(\mathcal{M}, \mathcal{V})$$

for each $n \in \mathbb{N}$.
(2) If $\mathcal{V}$ is a subspace of $B(H)$, and is a dual normal $\mathcal{M}$-module, then

$$H_{wcb}^n(\mathcal{M}, \mathcal{V}: /\mathcal{N}) = H_{wcb}^n(\mathcal{M}, \mathcal{V}).$$

We shall give the proof after some preliminary lemmas.

3.4.2 Lemma. *Let $\mathcal{N}$ be a hyperfinite von Neumann subalgebra of a von Neumann algebra $\mathcal{M}$ and let $\mathcal{V}$ be a dual normal $\mathcal{M}$-module. There is a continuous linear map $L_n: \mathcal{L}_w^n(\mathcal{M}, \mathcal{V}) \to \mathcal{L}_w^{n-1}(\mathcal{M}, \mathcal{V})$ such that $\phi - \partial L_n \phi$ is an $\mathcal{N}$-module map whenever $\phi \in \mathcal{L}_w^n(\mathcal{M}, \mathcal{V})$ with $\partial \phi = 0$. Further, the subscript "w" may be replaced by "wcb" on $\mathcal{L}$ throughout, and in both cases*

$$\|L_n\| \leq 2\gamma_n^2 + (n+1)\gamma_n \text{ where } \gamma_n = ((n+2)^n - 1))/(n+1).$$

Proof. Let $\mathcal{U}$ be an amenable subgroup of unitaries of $\mathcal{N}$ that generates $\mathcal{N}$ as a von Neumann algebra, and let $\mathcal{B}$ be the C^*-algebra generated by $\mathcal{U}$. The map K_n of Lemma 3.2.4 is restricted from $\mathcal{L}_c^n(\mathcal{M}, \mathcal{V})$ to $\mathcal{L}_w^n(\mathcal{M}, \mathcal{V})$ so

$$K_n: \mathcal{L}_w^n(\mathcal{M}, \mathcal{V}) \to \mathcal{L}_c^{n-1}(\mathcal{M}, \mathcal{V}),$$

with $\|K_n\| \leq ((n+2)^n - 1)/(n+1)$, such that if $\phi \in \mathcal{L}_w^n(\mathcal{M}, \mathcal{V})$ with $\partial \phi = 0$, then $\phi - \partial K_n \phi$ is a $\mathcal{B}$-module map. Now by Lemma 3.3.4 there is a continuous linear map $J_n\colon \mathcal{L}_c^n(\mathcal{M},\mathcal{V}:/\mathcal{B}) \to \mathcal{L}_c^{n-1}(\mathcal{M},\mathcal{V}:/\mathcal{B})$ such that if $\psi \in \mathcal{L}_c^n(\mathcal{M},\mathcal{V}:/\mathcal{B})$ with $\partial\psi = 0$, then $\psi - \partial J_n\psi$ is a normal map. With ϕ as above, $(I - \partial J_n)(I - \partial K_n)\phi$ is in $\mathcal{L}_w^n(\mathcal{M},\mathcal{V}:/\mathcal{B})$. Since $\mathcal{B}$ is weakly dense in $\mathcal{N}$, it follows that $(I - \partial J_n)(I - \partial K_n)\phi$ is an $\mathcal{N}$-module map. The lemma follows with

$$L_n = J_n + K_n - J_n\partial K_n$$

since

$$\|K_n\| \le ((n+2)^n - 1)/(n+1), \quad \|J_n\| \le ((n+2)^n - 1))/(n+1)$$

and

$$\|\partial\| \le (n+1)$$

in the continuous and completely bounded cases.

3.4.3 Lemma. *Let $\mathcal{N}$ be a hyperfinite von Neumann subalgebra of a von Neumann algebra $\mathcal{M}$ and let $\mathcal{V}$ be a dual normal $\mathcal{M}$-module. There is a surjective continuous linear map $P_n\colon \mathcal{L}_w^n(\mathcal{W},\mathcal{V}) \to \mathcal{L}_w^n(\mathcal{M},\mathcal{V}:/\mathcal{N})$ such that $\partial P_{n-1} = P_n\partial$. Further, the subscript "w" may be replaced by "wcb" on $\mathcal{L}$ throughout, and in both cases*

$$\|P_n\| \le 1.$$

Proof. This result will follow directly from Lemmas 3.2.6 and 3.3.4. Let $\mathcal{U}$ be an amenable group of unitaries in $\mathcal{N}$ that generates $\mathcal{N}$ as a von Neumann algebra, and let $\mathcal{B}$ be the C^*-algebra generated by $\mathcal{U}$. By Lemma 3.2.6 there is a continuous linear projection Q_n from $\mathcal{L}^n(\mathcal{M},\mathcal{V})$ onto $\mathcal{L}^n(\mathcal{M},\mathcal{V}:/\mathcal{B})$ such that $\partial Q_{n-1} = \partial Q_n$ (the algebra $\mathcal{A}$ in (3.2.6) is taken to be $\mathcal{M}$). In the notation of Lemma 3.3.4 with $\mathcal{A} = \mathcal{M}$, the map S_nT_n is a projection from $\mathcal{L}^n(\mathcal{M},\mathcal{V}:/\mathcal{B})$ onto $\mathcal{L}_w^n(\mathcal{M},\mathcal{V}:/\mathcal{B})$ satisfying

$$\partial S_{n-1}T_{n-1} = S_nT_n\partial.$$

Let P_n be $S_nT_nQ_n$ restricted to $\mathcal{L}_w^n(\mathcal{M},\mathcal{V})$. Then P_n is a projection from $\mathcal{L}_w^n(\mathcal{M},\mathcal{V})$ onto $\mathcal{L}_w^n(\mathcal{M},\mathcal{V}:/\mathcal{N})$ because

$$\mathcal{L}_w^n(\mathcal{M},\mathcal{V}:/\mathcal{N}) = \mathcal{L}_w^n(\mathcal{M},\mathcal{V}:/\mathcal{B})$$

since $\mathcal{B}$ is ultraweakly dense in $\mathcal{N}$. Further, $\partial P_{n-1} = P_n\partial$. The subscript "w" may be replaced throughout by "wcb" as this holds in each of the lemmas used. In both the continuous and completely bounded cases $\|P_n\| \le 1$ as this holds for S_n, T_n, Q_n. This proves the lemma.

3.4.4 Proof of Theorem 3.4.1. The embedding of $\mathcal{L}_w^n(\mathcal{M},\mathcal{V}:/\mathcal{N})$ into $\mathcal{L}_w^n(\mathcal{M},\mathcal{V})$ induces a linear map from $H_w^n(\mathcal{M},\mathcal{V}:/\mathcal{N})$ into $H_w^n(\mathcal{M},\mathcal{V})$. This map is injective by Lemma 3.4.3 and surjective by Lemma 3.4.2. The same conclusions hold in the completely bounded case.

3.4.5 Remark. Observe that if $\mathcal{L}$ is a von Neumann subalgebra of $\mathcal{M}$ with $\mathcal{L} \subseteq \mathcal{M} \cap \mathcal{N}'$, then the following cohomology groups are isomorphic:

$$H_w^n(\mathcal{M}, \mathcal{V} : /\mathcal{L}) \cong H_w^n(\mathcal{M}, \mathcal{V} : /(\mathcal{L} \cup \mathcal{N})'')$$

and

$$H_{wcb}^n(\mathcal{M}, \mathcal{V} : /\mathcal{L}) \cong H_{wcb}^n(\mathcal{M}, \mathcal{V} : /(\mathcal{L} \cup \mathcal{N})'').$$

3.4.6 Corollary. *If $\mathcal{M}$ is a hyperfinite von Neumann algebra and $\mathcal{V}$ is a dual normal $\mathcal{M}$-module, then*

$$H_c^n(\mathcal{M}, \mathcal{V}) = H_w^n(\mathcal{M}, \mathcal{V}) = H_{cb}^n(\mathcal{M}, \mathcal{V}) = H_{wcb}^n(\mathcal{M}, \mathcal{V}) = 0.$$

Proof. By Theorems 3.3.1 and 3.4.1 we need only show that

$$H_w^n(\mathcal{M}, \mathcal{V} : /\mathcal{M}) = 0, \quad H_{wcb}^n(\mathcal{M}, \mathcal{V} : /\mathcal{M}) = 0.$$

First consider an $\mathcal{M}$-multimodular cocycle $\phi \in \mathcal{L}^n(\mathcal{M}, \mathcal{V})$, and let $v_0 = \phi(1, \ldots, 1)$. Then

$$\phi(x_1, \ldots, x_n) = v_0 x_1 \ldots x_n = x_1 \ldots x_n v_0, \quad x_1, \ldots, x_n \in \mathcal{M},$$

by multimodularity. By letting $n - 1$ of the variables be 1, it follows that $x v_0 = v_0 x$ for $x \in \mathcal{M}$. If n is odd, then the cocycle condition $\partial\phi(1, \ldots, 1) = 0$ shows that $v_0 = 0$ and there is nothing to prove. If n is even, then define $\psi \in \mathcal{L}^{n-1}(\mathcal{M}, \mathcal{V})$ by

$$\psi(x_1, \ldots, x_{n-1}) = v_0 x_1 \ldots x_{n-1}, \quad x_1, \ldots, x_{n-1} \in \mathcal{M}.$$

It is then easy to check that $\phi = \partial\psi$. These maps are both completely bounded, and so this argument also settles the completely bounded case.

3.5 Notes and Remarks

The results in this section were proved in the continuous case by Johnson, Kadison and Ringrose [JKR], [KR2], [KR3]. Sketches of the modifications required in the completely bounded situation are in [ChES], [ChS4]. The presentation of these results here follows Ringrose's two surveys [Ri3,6] closely, and we wish to acknowledge our indebtedness to him.

Craw has given a proof of these results in the continuous case based on a more axiomatic approach to cohomology [Cr1,2]. His approach is not much simpler or shorter than that given here and depends crucially on turning tensor products and spaces of operators into modules. This leads to difficult technical problems in the completely bounded situation as we would become involved in the operator space versions of these constructions. The relevant constructions of tensors and turning spaces of completely bounded operators into operator spaces have been given by Blecher and Paulsen, and Effros and Ruan [BP], [ER1]. Given that the tensors and duals would have to be matched up, this approach is technically very complicated and we have not used it.

4

Completely Bounded Cohomology

4.1 Introduction

Completely bounded cohomology from a von Neumann algebra $\mathcal{M}$ into $B(H)$, or into a von Neumann algebra $\mathcal{N}$ containing $\mathcal{M}$ is shown to be zero in two cases in this chapter: when $\mathcal{N}$ is hyperfinite and when $\mathcal{N} = \mathcal{M}$.

The cohomology spaces $H_{cb}^n(\mathcal{M}, \mathcal{N})$ are shown to be zero in Theorem 4.2.6 for $\mathcal{N}$ hyperfinite containing $\mathcal{M}$, by reducing the completely bounded cocycle to a product of commutators and a single operator. The corresponding coboundary is then defined in terms of a projection onto a suitable closed subspace associated with a commutator. Here, and in the calculations, a commutator is taken to mean $[x, T] = x \cdot T - T \cdot x$, where x is in $\mathcal{M}$ and T is in some $\mathcal{M}$-module. In this chapter the modules will be spaces of operators with the module action arising from multiplication by $\theta(x)$ for some representation θ of $\mathcal{M}$. The technical Lemma 4.2.5 summarizes the reduction of a completely bounded cocycle from a von Neumann algebra into $B(H)$.

The projection ρ from $\mathcal{L}_{cb}(B(H), \mathcal{M})$ onto $\mathcal{L}_{cb}(B(H), \mathcal{M})_\mathcal{M}$, constructed in Section 1.8, is used to show that $H_{cb}^n(\mathcal{M}, \mathcal{M}) = 0$ (Theorem 4.3.1).

4.2 Completely Bounded Cohomology into $B(H)$

This section studies completely bounded cohomology from a von Neumann algebra $\mathcal{M}$ into $B(H)$, where $\mathcal{M}$ acts on the Hilbert space H. The methods apply to C^*-algebras just as easily, so the lemmas will be proved in this generality.

4.2.1 Lemma. *Let $\mathcal{A}$ be a C^*-algebra with $*$-representations θ and π on Hilbert spaces H and K, respectively. Let $T \in B(H, K)$ and let*

$$[x, T] = \pi(x)T - T\theta(x)$$

for all $x \in \mathcal{A}$. If q is the least projection in $B(K)$ such that

$$q[x, T] = [x, T]$$

for all $x \in \mathcal{A}$, then $[x, (1 - q)T] = 0$ and $q \in \pi(\mathcal{A})'$.

Proof. By definition, q is the projection from K onto the closure of the linear space

$$W = \{[x, T]\xi \colon \xi \in H, x \in \mathcal{A}\}.$$

Now W is invariant under $\pi(\mathcal{A})$ since

$$\pi(a)[x, T]\xi = (\pi(ax)T - T\theta(ax))\xi + (T\theta(a) - \pi(a)T)\theta(x)\xi$$
$$= [ax, T]\xi - [a, T]\theta(x)\xi \in W.$$

Since $\pi(\mathcal{A})$ is a $*$-subalgebra of $B(K)$ it follows that $\pi(a)q = q\pi(a)$ for all $a \in \mathcal{A}$. Hence

$$[a, (1 - q)T] = \pi(a)(1 - q)T - (1 - q)T\theta(a)$$
$$= (1 - q)[a, T]$$
$$= 0$$

for all $a \in \mathcal{A}$.

Let $\mathcal{A}$ be a C^*-algebra with $*$-representations θ and π on Hilbert spaces H and K, respectively. Regard $B(H, K)$ as a Banach $\mathcal{A}$-bimodule under the operation

$$x \cdot T \cdot y = \pi(x)T\theta(y)$$

for all $x, y \in \mathcal{A}$. Note that if $\mathcal{M}$ is a von Neumann algebra, and θ and π are normal representations then $B(H, K)$ is a dual normal $\mathcal{M}$-bimodule.

The following technical lemma concerning the representation of completely bounded maps is useful in simplifying subsequent calculations.

4.2.2 Lemma. *Let $\mathcal{A}$ be a unital C^*-algebra with unital C^*-subalgebra $\mathcal{B}$, and let θ and π be $*$-representations of $\mathcal{A}$ on Hilbert spaces H and K, respectively. Let $\phi \in \mathcal{L}_{cb}^n(\mathcal{A}, B(H, K))$ satisfy $\phi(x_1, \ldots, x_n) = 0$ whenever at least one argument is 1. Moreover, assume that ϕ is $\mathcal{B}$-multimodular. Then there are $*$-representations $\theta_1, \ldots, \theta_n$ of $\mathcal{A}$ on Hilbert spaces $H_1, \ldots, H_n$ and continuous linear operators*

$$t_j \colon H_{j+1} \to H_j \quad (0 \le j \le n),$$

where $H_0 = H$ and $H_{n+1} = K$, such that
(1) $\phi(x_1, \ldots, x_n) = t_0\theta_1(x_1)t_1 \ldots t_{n-1}\theta_n(x_n)t_n$ for all $x_j \in \mathcal{A}$,

(2) $\|\phi\|_{cb} = \|t_0\| \ldots \|t_n\|$,

(3) $t_{j-1}t_j = 0$ $(1 \le j \le n)$

(4) $\theta_j(x)t_j - t_j\theta_{j+1}(x) = [x, t_j] = 0$ for all $x \in \mathcal{B}$, and $0 \le j \le n$, where $\theta_0 = \theta$ and $\theta_{n+1} = \pi$, and

(5) if $\mathcal{A}$ is a von Neumann algebra, and ϕ is in $\mathcal{L}_{wcb}^n(\mathcal{A}, B(H, K))$ then the representations θ_j $(1 \le j \le n)$ may also be chosen to be normal.

Proof. Assume that $\|\phi\|_{cb} = 1$. By the representation theorem for completely bounded multilinear maps (Theorem 1.5.6), there are representations

$\theta_1, \ldots, \theta_n$ and operators $T_0, \ldots, T_n$ satisfying the conclusions of the lemma with the possible exception of

$$T_{j-1}T_j = 0, \qquad (1 \leq j \leq n).$$

Further, if ϕ is normal, then $\theta_1, \ldots, \theta_n$ may be chosen to be normal (Theorem 1.5.8). The proof now consists of modifying the operators $T_0, \ldots, T_n$ by multiplying each by suitable projections so that they satisfy the required additional relation.

There are two methods of making this choice: either by choosing a minimal family of projections by Zorn's Lemma, or by choosing projections onto certain subspaces of H_j inductively. The minimal family proof is given first, then the inductive proof.

To simplify the notation the representations θ_j $(1 \leq j \leq n)$ will be omitted and it will be assumed that x_j acts on H_j by

$$x_j \xi_j = \theta_j(x_j)\xi_j.$$

By Zorn's Lemma there is a minimal n-tuple of projections $(e_1, \ldots, e_n)$ in the commutants $\theta_j(\mathcal{A})'$ of the given representations θ_j such that

$$\phi(x_1, \ldots, x_n) = T_0 e_1 x_1 T_1 e_2 x_2 \ldots T_{n-1} e_n x_n T_n$$

for all x_j. Replacing T_0 by $t_0 = T_0 e_1$, T_j by $t_j = e_j T_j e_{j+1}$ $(1 \leq j \leq n-1)$ and T_n by $t_n = e_n T_n$ implies that we may assume that the subspaces spanned by the sets

$$\{x_{j+1}t_{j+1} \ldots x_n t_n \xi \colon x_k \in \mathcal{A}, \xi \in H\} \text{ and}$$
$$\{x_{j-1}^* t_{j-2}^* \ldots x_1^* t_0^* \eta \colon x_k \in \mathcal{A}, \eta \in K\}$$

are dense in H_{j+1} and H_{j-1}, respectively. To see this, let f_{j+1} and g_{j-1} be the projections onto the closures of these two subspaces. Then we obtain a valid n-tuple

$$(e_1, \ldots, e_{j-2}, g_{j-1}, e_j, f_{j+1}, e_{j+2}, \ldots, e_n)$$

with $g_{j-1} \leq e_{j-1}$ and $f_{j+1} \leq e_{j+1}$; the minimality of the n-tuple $(e_1, \ldots, e_n)$ ensures that $g_{j-1} = e_{j-1}$ and $f_{j+1} = e_{j+1}$. Now

$$\langle \phi(x_1, \ldots, x_{j-1}, 1, x_{j+1}, \ldots, x_n)\xi, \eta \rangle$$
$$= \langle t_{j-1}t_j x_{j+1}t_{j+1} \ldots x_n t_n \xi, x_{j-1}^* t_{j-2}^* \ldots x_1^* t_0^* \eta \rangle$$

for all x_k in $\mathcal{A}$, ξ in H and η in K. Hence $t_{j-1}t_j = 0$, proving (3).

If $b \in \mathcal{B}$ then, for $\xi \in H$, $\eta \in K$,

$$0 = \langle (\phi(x_1, \ldots, x_j b, x_{j+1}, \ldots, x_n) - \phi(x_1, \ldots, x_j, bx_{j+1}, \ldots, x_n))\xi, \eta \rangle$$
$$= \langle (bt_j - t_j b)x_{j+1}t_{j+1} \ldots x_n t_n \xi, x_j^* t_{j-1}^* \ldots x_1^* t_0^* \eta \rangle,$$

and so the commutator $[bt_j - t_j b]$ is 0. This completes the proof of the lemma, since (1) and (2) are clear from the constructions of the t_j's.

4.2.3 Alternative Proof of Lemma 4.2.2

Here is an alternative proof of Lemma 4.2.2. The first stage of the proof is the same.

For $0 \le j \le n-1$, let q_j be the projection from H_{j+1} onto the closure of the subspace

$$\{x_{j+1}T_{j+1}\ldots x_n T_n \xi : x_k \in \mathcal{A}, \quad (j+1 \le k \le n), \; \xi \in H\}.$$

Then $q_j \in \theta_{j+1}(\mathcal{A})'$ and q_j is the least projection in $B(H_{j+1})$ such that

$$(1 - q_j)x_{j+1}T_{j+1}\ldots T_{n-1}x_n T_n = 0.$$

The projections $p_1, \ldots, p_n$ are defined inductively as follows. Let p_1 be the projection from H_1 onto the closed linear span of

$$\{x_1 q_0 T_0^* \xi : x_1 \in \mathcal{A} \quad \text{and} \quad \xi \in H\}.$$

If $p_1, \ldots, p_{j-1}$ have already been chosen, define p_j to be the projection from H_j onto the closed linear span of

$$\{x_j q_{j-1} T_{j-1}^* p_{j-1} x_{j-1} \ldots T_2^* p_2 x_2 q_1 T_1^* p_1 x_1 q_0 T_0 \xi : x_k \in \mathcal{A} \, (1 \le k \le j), \, \xi \in H\}.$$

Then $p_j \in \theta_j(\mathcal{A})'$ and p_j is the least projection such that

$$T_0 q_0 x_1 p_1 T_1 q_1 \ldots p_{j-1}T_{j-1}q_{j-1}x_j(1 - p_j) = 0.$$

By induction on j, it follows that

(1) $\phi(x_1, \ldots, x_n) = T_0 q_0 x_1 p_1 T_1 q_1 \ldots p_{j-1}T_{j-1}q_{j-1}x_j T_j x_{j+1} \ldots x_n T_n$

$\qquad\qquad = T_0 q_0 x_1 p_1 T_1 q_1 \ldots p_{j-1}T_{j-1}q_{j-1}x_j p_j T_j x_{j+1} \ldots x_n T_n$

for all j. This leads to

(2) $\qquad \phi(x_1, \ldots, x_n) = T_0 q_0 x_1 p_1 T_1 q_1 \ldots p_{n-1}T_{n-1}q_{n-1}x_n p_n T_n.$

In passing from the second equality in (1) with j to the first with $j+1$ the definition of q_j as the least projection with

$$q_j x_{j+1} T_{j+1} \ldots T_{n-1} x_n T_n = x_{j+1}T_{j+1} \ldots T_{n-1} x_n T_n$$

is used, multiplying by the remainder of the product on the left. To pass from the first equality with j to the second with j, the definition of p_j is used via

$$T_0 q_0 x_1 p_1 T_1 q_1 \ldots p_{j-1} T_{j-1} q_{j-1} x_j p_j = T_0 q_0 x_1 p_1 T_1 q_1 \ldots p_{j-1} T_{j-1} q_{j-1} x_j,$$

multiplying on the right by the remainder of the product. The operators t_j required in the conclusion are now just $T_0 q_0$, $p_j T_j q_j$ $(1 \le j \le n-1)$, and $p_n T_n$. For these operators,

$$\|t_j\| = \|p_j T_j q_j\| \le \|T_j\|$$

and so a representation of ϕ is obtained which satisfies

$$\|\phi\|_{cb} \ge \|T_0 q_0\| \cdot \|p_1 T_1 q_1\| \ldots \|p_{n-1} T_{n-1} q_{n-1}\| \cdot \|p_n T_n\|.$$

The representation theorem for completely bounded multilinear maps (1.5.6) implies that equality holds here. It only remains to prove that

$$p_{j-1} T_{j-1} q_{j-1} p_j T_j q_j = 0.$$

The equation

$$\phi(x_1, \ldots, x_{j-1}, 1, x_{j+1}, \ldots, x_n) = 0$$

for all $x_k \in \mathcal{A}$ $(k \ne j)$, together with the representation (1), shows that

$$p_0 T_0 q_0 x_1 p_1 T_1 q_1 \ldots x_{j-1} p_{j-1} T_{j-1} q_{j-1} p_j T_j x_{j+1} T_{j+1} \ldots x_n T_n = 0.$$

Using the definitions of p_{j-1} and q_j as the projections onto the subspaces associated with them, leads from this equality to

$$p_{j-1} p_{j-1} T_{j-1} q_{j-1} p_j T_j q_j = 0$$

as required, since p_{j-1} is a projection.

It remains to prove (4) in (4.2.2). Note that the equation

$$\phi(x_1, \ldots, x_j b, x_{j+1}, \ldots, x_n) = \phi(x_1, \ldots, x_j, b x_{j+1}, \ldots, x_n)$$

implies that

$$T_0 q_0 x_1 p_1 \ldots T_{j-1} q_{j-1} x_j (b p_j T_j q_j - p_j T_j q_j b) x_{j+1} p_{j+1} T_{j+1} \ldots T_n = 0$$

for all $x_k \in \mathcal{A}$ and $b \in \mathcal{B}$. The definition of the projections p_j and q_j yields

$$p_j (b p_j T_j q_j - p_j T_j q_j b) q_j = 0$$

for all $b \in \mathcal{B}$. The equation required in (4) follows since

$$p_j \in \theta_j(\mathcal{A})' \quad \text{and} \quad q_j \in \theta_{j+1}(\mathcal{A})'.$$

The following algebraic lemma is useful in the subsequent calculations.

4.2.4 Lemma. *Let ϕ be a completely bounded n-linear map from a C^*-algebra $\mathcal{A}$ into $B(H, K)$ with a representation*

$$\phi(x_1,\ldots,x_n) = T_0\theta_1(x_1)T_1 \ldots T_{n-1}\theta_n(x_n)T_n,$$

where the θ_j's are $*$-representations on Hilbert spaces, and the T_j's are bridging operators. Let

$$[x, T_j] = \theta_j(x)T_j - T_j\theta_{j+1}(x)$$

for all j.
  (i) If
$$\psi(x_1,\ldots,x_n) = [x_1, T_0] \ldots [x_n, T_{n-1}]T_n,$$

  then
$$\partial\psi(x_0,\ldots,x_n) = (-1)^n[x_0, T_0] \ldots [x_n, T_n].$$

(ii) If $T_{j-1}T_j = 0$ for $1 \leq j \leq n$, then

$$\phi(x_1,\ldots,x_n) = (-1)^n[x_1, T_0] \ldots [x_n, T_{n-1}]T_n$$

  and
$$\partial\phi(x_0,\ldots,x_n) = [x_0, T_0] \ldots [x_n, T_n].$$

Proof. The representations will be suppressed in the proof, and $\theta_j(x)$ will just be written x; on which spaces it acts determines the representation θ_j.

(i) If an n-linear map γ is a product of an $(n-1)$-linear map α and a linear map β by

$$\gamma(x_1,\ldots,x_n) = \alpha(x_1,\ldots,x_{n-1})\beta(x_n)$$

then

$$\begin{aligned}
\partial\gamma(x_0,\ldots,x_n) = {} & x_0\alpha(x_1,\ldots,x_{n-1})\beta(x_n) \\
& + \sum_{j=1}^{n-1}(-1)^j\alpha(x_0,\ldots,x_{j-1}x_j,\ldots,x_{n-1})\beta(x_n) \\
& + (-1)^n\alpha(x_0,\ldots,x_{n-2})\beta(x_{n-1}x_n) \\
& + (-1)^{n+1}\alpha(x_0,\ldots,x_{n-2})\beta(x_{n-1})x_n.
\end{aligned}$$

Add and subtract $(-1)^n\alpha(x_0,\ldots,x_{n-2})x_{n-1}\beta(x_n)$ to obtain

$$\partial\gamma = (\partial\alpha)\beta + (-1)^{n+1}\alpha(\partial\beta).$$

In particular, if β is a derivation then

$$\partial\gamma = (\partial\alpha)\beta,$$

and this equation proves inductively that if γ is a product of n derivations then $\partial\gamma = 0$.

Now write $\psi = \gamma\beta$, where

$$\gamma(x_1, \ldots, x_{n-1}) = [x_1, T_0] \ldots [x_{n-1}, T_{n-2}], \quad \beta(x_n) = [x_n, T_{n-1}]T_n.$$

By a direct calculation

$$\partial\beta(x_{n-1}, x_n) = -[x_{n-1}, T_{n-1}][x_n, T_n].$$

Then the equation

$$\partial\psi(x_0, \ldots, x_n) = (-1)^n[x_0, T_0] \ldots [x_n, T_n]$$

is immediate from the relations

$$\partial\psi = (\partial\gamma)\beta + (-1)^{n+1}\gamma(\partial\beta), \quad \partial\gamma = 0.$$

(ii) This is a straightforward induction on the number of Lie products in the representation of ϕ using the relationship $T_{j-1}T_j = 0$ to insert the additional zero terms. Here are some of the steps:

$$
\begin{aligned}
\phi(x_1, \ldots, x_n) &= T_0 x_1 T_1 \ldots x_n T_n \\
&= (-1)[x_1, T_0]T_1 x_2 T_2 \ldots x_n T_n \\
&= (-1)^2[x_1, T_0][x_2, T_1]T_2 \ldots x_n T_n \\
&= \cdots\cdots \\
&= (-1)^n[x_1, T_0][x_2, T_1] \ldots [x_n, T_{n-1}]T_n
\end{aligned}
$$

as required.

The main technical result of this section is the following lemma, which easily leads to Theorem 4.2.6 and is used to prove Theorem 4.3.1.

4.2.5 Lemma. *Let $\mathcal{N} \subseteq \mathcal{M}$ be von Neumann subalgebras of $B(H)$, and let ϕ be a completely bounded $\mathcal{N}$-module map from $\mathcal{M}^n$ into $B(H)$ with $\partial\phi = 0$ and $\phi(x_1, \ldots, x_n) = 0$ if at least one $x_j = 1$ $(1 \leq j \leq n)$. Then there are representations $\theta_1, \ldots, \theta_{n-1}$ of $\mathcal{M}$ on Hilbert spaces $H_1, \ldots, H_{n-1}$ and continuous linear operators $t_j \colon H_j \to H_{j-1}$ $(1 \leq j \leq n)$, where $H_0 = H_n = H$, such that*

(1) if

$$\psi(x_1, \ldots, x_{n-1}) = (-1)^{n-1}[x_1, t_1] \ldots [x_{n-1}, t_{n-1}]t_n$$
$$= t_1 x_1 t_2 \ldots t_{n-1} x_{n-1} t_n$$

for all $x_j \in \mathcal{M}$, then

(2)
$$\phi(x_1, \ldots, x_n) = \partial \psi(x_1, \ldots, x_n)$$
$$= [x_1, t_1] \ldots [x_n, t_n]$$

for all $x_j \in \mathcal{M}$,

(3)
$$\|t_1\| \ldots \|t_n\| \le \|\phi\|_{cb},$$

(4)
$$t_j t_{j+1} = 0 \quad \text{for} \quad j = 1, \ldots, n-1,$$

(5)
$$[x, t_j] = \theta_{j-1}(x)t_j - t_j\theta_j(x) = 0$$

for all $x \in \mathcal{N}$ and $1 \le j \le n$ where $\theta_0 = \theta_{n+1}$ is the identity representation of $\mathcal{M}$ on H, and

(6) if ϕ is normal, then the representations θ_j can be chosen to be normal.

Proof. By Lemmas 4.2.2 and 4.2.4 (ii) ϕ has a representation of the form

(1)
$$\phi(x_1, \ldots, x_n) = [x_1, t_1][x_2, t_2] \ldots [x_n, t_n]t_{n+1}$$

for suitable representations θ_j, which are omitted, and suitable bridging operators t_j. Thus by Lemma 4.2.4 (ii),

(2)
$$(-1)^n \partial\phi(x_0, \ldots, x_n) = [x_0, t_1] \ldots [x_n, t_{n+1}] = 0$$

for all $x_j \in \mathcal{M}$. Let q be the least projection in $B(K_n)$ such that

$$q[x, t_{n+1}] = [x, t_{n+1}]$$

for all $x \in \mathcal{M}$, where K_n is the Hilbert space associated with θ_n. Here θ_j acts on H_j for $j < n$ but on K_n for $j = n$ as this space will be modified to form H_n of the conclusion. Then equation (2) implies that

(3)
$$[x_0, t_1] \ldots [x_{n-1}, t_n]q = 0$$

as q is the projection onto the closed linear span of

$$\{[x, t_{n+1}]\xi : \xi \in H, x \in \mathcal{M}\}.$$

Let ψ be the completely bounded $(n-1)$-linear map from $\mathcal{M}$ into $B(H)$ defined by

$$\psi(x_1,\ldots,x_{n-1}) = t_1 x_1 t_2 \ldots t_{n-1} x_{n-1} t_n (1-q) t_{n+1}$$

for all $x_j \in \mathcal{M}$. Let t_n of the conclusion be $t_n(1-q)t_{n+1} = -t_n q t_{n+1}$. Note that by Lemma 4.2.4,

$$\psi(x_1,\ldots,x_{n-1}) = [x_1,t_1]\ldots[x_{n-1},t_{n-1}]t_n(1-q)t_{n+1}$$

and that

$$\partial\psi(x_1,\ldots,x_n) = [x_1,t_1]\ldots[x_{n-1},t_{n-1}][x_n,t_n(1-q)t_{n+1}].$$

By the definition of q and Lemma 4.2.1, $[x,(1-q)t_{n+1}] = 0$ so that

$$[x_n,t_n(1-q)t_{n+1}] = [x_n,t_n](1-q)t_{n+1}.$$

Hence
$$\partial\psi(x_1,\ldots,x_n) = [x_1,t_1]\ldots[x_n,t_n](1-q)t_{n+1}$$
$$= \phi(x_1,\ldots,x_n)$$

by (1). The conclusions follow directly from this and Lemma 4.2.2.

The main theorem of this section follows by applying a conditional expectation to the map ψ of the previous lemma and using the module property of a conditional expectation.

4.2.6 Theorem. *If $\mathcal{A}$ is a C^*-subalgebra of a hyperfinite von Neumann algebra $\mathcal{N}$, then $H^n_{cb}(\mathcal{A},\mathcal{N}) = 0$ for all $n \in \mathbb{N}$.*

Proof. Regard $\mathcal{N}$ as a von Neumann subalgebra of $B(H)$ by taking a faithful normal representation of $\mathcal{N}$ on a Hilbert space H. The C^*-algebra $\mathcal{A}$ is then a C^*-subalgebra of $B(H)$ and $\mathcal{N}$ is a normal $\mathcal{A}$-bimodule. Hence

$$H^n_{cb}(\mathcal{A},\mathcal{N}) \cong H^n_{cb}(\mathcal{M},\mathcal{N}),$$

where $\mathcal{M} = \mathcal{A}''$ is the von Neumann algebra generated by $\mathcal{A}$ in $B(H)$, by Theorem 3.3.1. Note that the part of $\mathcal{N}$ annihilated by the identity of $\mathcal{M}$ plays no role in the cohomology calculation so we can assume the identity of $\mathcal{M}$ is the identity of $\mathcal{N}$. By Theorem 3.2.7

$$H^n_{cb}(\mathcal{M},\mathcal{N} : /\mathbb{C}1) \cong H^n_{cb}(\mathcal{M},\mathcal{N}).$$

If ϕ is a completely bounded n-linear map from $\mathcal{M}$ into $\mathcal{N}$ with

$$\partial\phi = 0 \quad \text{and} \quad \phi(x_1,\ldots,x_n) = 0$$

if $x_j = 1$ for some $1 \leq j \leq n$, then there exists $\psi \in \mathcal{L}_{cb}^{n-1}(\mathcal{M}, B(H))$ such that $\partial \psi = \phi$ by Lemma 4.2.5.

Let E be a conditional expectation from $B(H)$ onto $\mathcal{N}$, which exists by the injectivity of $\mathcal{N}$. Then E is $\mathcal{N}$-modular, and hence $\mathcal{M}$-modular, so

$$\partial(E\psi)(x_1, \ldots, x_n) = E\partial\psi(x_1, \ldots, x_n)$$
$$= \phi(x_1, \ldots, x_n) \quad \text{for all} \quad x_j \in \mathcal{A}$$

because ϕ maps into $\mathcal{N}$.

4.2.7 Remark. From the above result one may deduce a theorem for $\mathcal{A}$ having two representations on Hilbert spaces H and K and for the bimodule $B(H, K)$ in place of $\mathcal{N}$.

4.3 Completely Bounded Cohomology into the Algebra

Theorems 4.2.6 and 1.7.4 are combined in this short section to prove that the completely bounded (normal) Hochschild cohomology of a von Neumann algebra over itself is zero.

4.3.1 Theorem. *If $\mathcal{M}$ is a von Neumann algebra, then*

$$H_{cb}^n(\mathcal{M}, \mathcal{M}) = H_{wcb}^n(\mathcal{M}, \mathcal{M}) = 0.$$

Proof. By Theorem 3.3.1, it is sufficient to prove that either one of these cohomology groups is zero, as they are isomorphic. However the proof applies to either space of maps because the techniques of Section 4.2 apply to both the completely bounded and the normal completely bounded cases. The proof given here will be for the completely bounded situation; normality is handled by choosing all the ∗-representations to be normal.

Let $\phi \colon \mathcal{M}^n \to \mathcal{M}$ be a completely bounded n-cocycle. By Theorem 1.7.4 there exists a projection ρ from $\mathcal{L}_{cb}^1(\mathcal{M}, \mathcal{M})$ onto the space $\mathcal{L}_{cb}^1(\mathcal{M}, \mathcal{M})_{\mathcal{M}}$ of right $\mathcal{M}$-module maps in $\mathcal{L}_{cb}^1(\mathcal{M}, \mathcal{M})$, with the special properties outlined in that theorem. By fixing the first $n-1$ variables of an n-linear completely bounded map ξ, we may let ρ act on the last variable and we write the resulting action as $\rho\xi$. Then $\rho\xi$ is a completely bounded n-linear map from $\mathcal{M}^n$ to $\mathcal{M}$ which is $\mathcal{M}$-modular on the right in the last variable. From the construction of ρ in Theorem 1.7.4, there exists a net (m_j^α) of sequences from $\mathcal{M}$ satisfying

(i) $\sum_j m_j^{\alpha*} m_j^\alpha = 1$ for all α,

(ii) $\rho\xi(x_1, \ldots, x_n)$ is the ultraweak limit over α of $\sum_j \xi(x_1, \ldots, x_{n-1}, x_n m_j^{\alpha*}) m_j^\alpha$

   for all $x_1, \ldots, x_n \in \mathcal{M}$.

Now let

$$\psi(x_1, \ldots, x_{n-1}) = (-1)^n \rho \phi(x_1, \ldots, x_{n-1}, 1)$$

for all $x_1, \ldots, x_{n-1} \in \mathcal{M}$. The cocycle equation

$$\partial \phi(x_1, \ldots, x_n, m_j^{\alpha*}) m_j^\alpha = 0$$

implies that

$$x_1 \sum_j \phi(x_2, \ldots, x_n, 1 m_j^{\alpha*}) m_j^\alpha$$

$$+ \sum_{k=1}^{n-1} (-1)^k \sum_j \phi(x_1, \ldots, x_k x_{k+1}, \ldots, x_n, 1 m_j^{\alpha*}) m_j^\alpha$$

$$+ (-1)^n \sum_j \phi(x_1, \ldots, x_{n-1}, x_n m_j^{\alpha*}) m_j^\alpha$$

$$+ (-1)^{n+1} \sum_j \phi(x_1, \ldots, x_n) m_j^{\alpha*} m_j^\alpha$$

$$= 0$$

for all α. Taking ultraweak limits over α and using (i) and (ii) above leads to

$$x_1 \rho \phi(x_2, \ldots, x_n, 1) + \sum_{k=1}^{n-1} (-1)^k \rho \phi(x_1, \ldots, x_k x_{k+1}, \ldots, x_n, 1)$$

$$+ (-1)^n \rho \phi(x_1, \ldots, x_n) + (-1)^{n+1} \phi(x_1, \ldots, x_n)$$

$$= 0.$$

Thus, from this last equation,

$$\partial \rho \phi(x_1, \ldots, x_n) + (-1)^{n+1} \phi(x_1, \ldots, x_n) = 0$$

since

$$\rho \phi(x_1, \ldots, x_n) = \rho \phi(x_1, \ldots, x_{n-1}, 1) x_n$$

by the construction of ρ. We have proved that $\phi = \partial \psi$, where $\|\psi\|_{cb} \le \|\phi\|_{cb}$, and so every completely bounded n-cocycle is a coboundary. It follows that $H^n_{cb}(\mathcal{M}, \mathcal{M}) = 0$.

4.4 Notes and Remarks

The results of Section 4.2 are taken from Section 4 of [ChES]. Completely bounded multilinear maps were originally introduced in [ChS1] partly with a view to using them in cohomology calculations. There is an error in the inductive choice of the sequence p_j of projections in the proof of Lemma 4.2 of [ChES]. This error was noticed by a student of U. Haagerup, and is corrected in the appendix of [ChS4]. Two proofs are given of this step in Lemma 4.2.2: a maximal family proof due to U. Haagerup and a correct inductive construction of p_j and q_j. Section 4.3 is from [ChS5].

5

Hyperfinite Subalgebras

5.1 Introduction

This chapter contains three different techniques, each associated with hyperfinite von Neumann algebras, which play an important role in the next chapter on continuous cohomology. The technique in the first section is due to Kadison and Ringrose [KR3] and is based on a generalization of the old idea of Murray and von Neumann that if $x \in \mathcal{M}$ and $y \in \mathcal{M}'$ with $xy = 0$ then there is a $z \in \mathcal{M} \cap \mathcal{M}'$, the centre of both $\mathcal{M}$ and $\mathcal{M}'$, such that $xz = 0$ and $zy = y$. For module maps over the centre this enables us to extend multilinear maps algebraically by a subalgebra of the commutant, provided this subalgebra is abelian. The extension can be shown to be norm continuous and behaves well with respect to the coboundary operator. This theory is developed in Section 5.2 and leads to the continuous cocycles from $\mathcal{M}^n$ into $\mathcal{M}$ being coboundaries into larger algebras.

Section 5.3 is devoted to three remarkable results of Popa on the existence of hyperfinite subfactors in type II_1 factors with trivial relative commutants, and a result of his on maximal abelian self-adjoint subalgebras (masas). The first is that there is a hyperfinite subfactor $\mathcal{N}$ of a type II_1 factor $\mathcal{M}$ with trivial relative commutant (that is, $\mathcal{N}' \cap \mathcal{M} = \mathbb{C}1$) (5.3.6). In the second $\mathcal{N}$ is constructed to contain a preassigned Cartan subalgebra $\mathcal{A}$ (5.3.9). In the third a masa in $B(L^2(\mathcal{M}))$ is constructed from a Cartan subalgebra (5.3.11).

Section 5.4 is the result of combining Popa's results of Section 5.3 and the non-commutative Haagerup–Pisier–Grothendieck inequality with an averaging argument. The averaging removes the anti-isomorphism term in the representation of the bilinear form from the above inequality. This is crucial in showing that suitable bilinear module maps lift to the Haagerup tensor product (5.4.6). The lift to the tensor product, combined with results of Section 1.6, shows the maps to be completely bounded.

5.2 Extension by a Masa

In this section the multilinear maps arising in cohomology are lifted to a larger algebra generated by the von Neumann algebra and an abelian von Neumann algebra in the commutant. When this abelian algebra is a maximal abelian ∗-subalgebra of the commutant, then cocycles with images in the algebra can be shown to be coboundaries into a larger algebra (5.2.3 and 5.2.4).

5.2.1 Remarks. Before the results of this section are stated and proved some von Neumann algebra notation and results are recalled.

Let $\mathcal{M}$ be a von Neumann algebra with centre $\mathcal{Z}(= \mathcal{M} \cap \mathcal{M}')$. The following well-known result traces its origin back to the first paper of Murray and von Neumann [MvN1] [Di2, Proposition 7, p. 27], [KR4, Theorem 5.5.4, p. 333].

If $x_1, \ldots, x_n \in \mathcal{M}$ and $y_1, \ldots, y_n \in \mathcal{M}'$ with

$$\sum_1^n x_j y_j = 0,$$

then there is an $n \times n$ matrix $(c_{ij}) \in \mathbf{M}_n(\mathcal{Z})$ such that

$$(x_i)(c_{ij}) = 0 \quad \text{and}$$
$$(c_{ij})(y_j)^T = (y_j)^T$$

where $(x_i) = (x_1, \ldots, x_n)$ and $(y_j)^T$ is the corresponding column of y's.

5.2.2 Lemma. *Let $\mathcal{M}$ be a von Neumann algebra on a Hilbert space H with centre $\mathcal{Z}$, let $\mathcal{B}$ be an abelian von Neumann algebra in $\mathcal{M}'$ and let $C^*(\mathcal{M}, \mathcal{B})$ denote the C^*-algebra generated by $\mathcal{M}$ and $\mathcal{B}$.*

(1) If $\mathcal{V}$ is a subspace of $\mathcal{B}'$ and an $\mathcal{M}$-module, and $(\mathcal{B}\mathcal{V})^-$ is the Banach $\mathcal{B}$-module generated by $\mathcal{B}$ and $\mathcal{V}$, then there is an isometric embedding $I_\mathcal{B}$ of $\mathcal{L}^n(\mathcal{M}, \mathcal{V}:/\mathcal{Z})$ into $\mathcal{L}^n(C^*(\mathcal{M}, \mathcal{B}), (\mathcal{B}\mathcal{V})^-:/\mathcal{B})$ given by

$$(I_\mathcal{B}\phi)(b_1 x_1, \ldots, b_n x_n) = b_1 \ldots b_n \phi(x_1, \ldots, x_n)$$

for all $b_1, \ldots, b_n \in \mathcal{B}$ and $x_1, \ldots, x_n \in \mathcal{M}$. Further, $\partial I_\mathcal{B} = I_\mathcal{B} \partial$.

(2) If $\mathcal{A}$ is a C^*-subalgebra of $\mathcal{M}$ containing $\mathcal{Z}$, then $I_\mathcal{B}$ maps

$$\mathcal{L}^n(\mathcal{M}, \mathcal{V}:/\mathcal{A}) \text{ into } \mathcal{L}^n(C^*(\mathcal{M}, \mathcal{B}), (\mathcal{B}\mathcal{V})^-:/C^*(\mathcal{A}, \mathcal{B})).$$

Proof. Note that $\mathcal{B}$ commutes with $\mathcal{V}$ and $\mathcal{M}$ so that $(\mathcal{B}\mathcal{V})^-$ is a $C^*(\mathcal{M}, \mathcal{B})$-module. Let $\mathcal{P}$ denote the lattice of projections in $\mathcal{B}$ and let $\mathcal{B}_0$ denote the algebra of all operators of the form $p_1 x_1 + \cdots + p_m x_m$ where $p_1, \ldots, p_m \in \mathcal{P}$ and $x_1, \ldots, x_m \in \mathcal{M}$. Then the norm closure of $\mathcal{B}_0$ is $C^*(\mathcal{M}, \mathcal{B})$, since $\mathcal{B}_0 \subseteq C^*(\mathcal{M}, \mathcal{B})$ and the norm closure of $\mathcal{B}_0$ contains $\mathcal{M}$ and $\mathcal{B}$.

Let $\phi \in \mathcal{L}^n(\mathcal{M}, \mathcal{V}:/\mathcal{Z})$. The map $I_\mathcal{B}\phi$ is defined by the $\mathcal{B}$-module property initially on $\mathcal{B}_0$, is shown to be continuous and then lifted to $C^*(\mathcal{M}, \mathcal{B})$. If $y_1, \ldots, y_n$ are in $\mathcal{B}_0$ with

$$y_j = \sum_{k=1}^{m(j)} p_{j,k} x_{j,k},$$

where $p_{j,k} \in \mathcal{P}$ and $x_{j,k} \in \mathcal{M}$ for all j, k, then define

(1) $(I_\mathcal{B}\phi)(y_1, \ldots, y_n) =$

$$\sum_{k(1)=1}^{m(1)} \cdots \sum_{k(n)=1}^{m(n)} p_{1,k(1)} \cdots p_{n,k(n)} \phi(x_{1,k(1)}, \ldots, x_{n,k(n)}).$$

To show that the map $I_B\phi$ is well defined it is sufficient to show that if

$$y_\ell = \sum_{k(\ell)=1}^{m(\ell)} p_{\ell,k(\ell)} x_{\ell,k(\ell)} = 0$$

for some ℓ, then

$$(I_B\phi)(y_1,\ldots,y_n) = 0.$$

Now by the relation between a von Neumann algebra and its commutant discussed before this lemma, there is an $m(\ell) \times m(\ell)$ matrix (c_{ij}) with $c_{ij} \in \mathcal{Z}$ such that

(2) $\quad (I - (c_{ij}))(p_{\ell,1},\ldots,p_{\ell,m(\ell)})^T = 0$ and $(x_1,\ldots,x_{\ell,m(\ell)})(c_{ij}) = 0.$

    Now
$$\sum_{k(\ell)=1}^{m(\ell)} p_{\ell,k(\ell)}\phi(x_{1,k(1)},\ldots,x_{n,k(n)})$$
$$= \sum_{k(\ell)}\sum_{j} c_{k(\ell),j} p_{\ell,j}\phi(x_{1,k_1},\ldots,x_{n,k(n)})$$
$$= \sum_{j} p_{\ell,j}\phi(x_{1,k_1},\ldots,\sum_{k(\ell)} c_{k(\ell),j} x_{\ell,k(\ell)},\ldots,x_{n,k(n)})$$
$$= 0$$

by (2) and the $\mathcal{Z}$-module property of ϕ.

    To show that I_B is an isometry the elements y_j in $\mathcal{B}_0$ are written in terms of the same projections just by considering the finite dimensional algebra generated by $\{p_{i,j}:$ all $i,j\}$ associated with (1). Letting q_j ($1 \leq j \leq m$) denote the minimal projections of this finite dimensional algebra, each y_i may be expressed as a sum

$$y_i = \sum_{j} q_j w_{ij} = \sum_{j} q_j w_{ij} q_j$$

for $1 \leq i \leq n$, where $w_{ij} \in \mathcal{M}$. Now

(3) $$\|y_i\| = \max\{\|w_{ij}\|: 1 \leq j \leq m\}$$

because $\{q_j\}$ is a set of orthogonal projections. Equation (1) reduces, by the orthogonality of the q_j's, to

$$(I_B(\phi)(y_1,\ldots,y_n) = \sum_{j} q_j\phi(w_{1j},\ldots,w_{nj})$$
$$= \sum_{j} q_j\phi(w_{1j},\ldots,w_{nj})q_j$$

because $\mathcal{V} \subseteq \mathcal{B}'$. Hence

$$\|(I_\mathcal{B}\phi)(y_1,\ldots,y_n)\| = \max\{\|\phi(w_{1j},\ldots,w_{nj})\|\colon 1 \le j \le m\}$$
$$\le \|\phi\|\,\|y_1\|\cdots\|y_n\|$$

by (3). Hence $\|I_\mathcal{B}\phi\| = \|\phi\|$ on $\mathcal{B}_0$.

Thus $I_\mathcal{B}\phi$ extends by norm continuity to a map on $C^*(\mathcal{M},\mathcal{B})^n$ into $(\mathcal{B}\mathcal{V})^-$. Since $\mathcal{B}$ is the norm closed span of its projections, $\mathcal{B}$-modularity follows from continuity and equation (1). The equation

$$I_\mathcal{B}\partial = \partial I_\mathcal{B}$$

follows directly from the definition (1) because $\mathcal{V}$ is contained in $\mathcal{B}'$. Modularity with respect to $\mathcal{A}$ and $C^*(\mathcal{A},\mathcal{B})$ is immediate from the definition of $I_\mathcal{B}\phi$. This proves the lemma.

Observe that if $\mathcal{B}$ is a masa in $\mathcal{M}'$, then the von Neumann algebra $(\mathcal{M} \cup \mathcal{B})''$ generated by $\mathcal{M}$ and $\mathcal{B}$ is equal to $\mathcal{B}'$. This is because

$$(\mathcal{M} \cup \mathcal{B})' = \mathcal{M}' \cap \mathcal{B}' = \mathcal{B},$$

since $\mathcal{B}$ is a masa in $\mathcal{M}'$. Hence the C^*-algebra $C^*(\mathcal{M},\mathcal{B})$ is ultraweakly dense in $\mathcal{B}'$, and $\mathcal{L}^n(C^*(\mathcal{M},\mathcal{B}), C^*(\mathcal{M},\mathcal{B}))$ is contained in $\mathcal{L}^n(C^*(\mathcal{M},\mathcal{B}), \mathcal{B}')$. The embedding $I_\mathcal{B}$ of Lemma 5.2.2 is regarded as going into this latter space.

5.2.3 Proposition. *Let $\mathcal{M}$ be a von Neumann algebra with centre $\mathcal{Z}$ and C^*-subalgebra $\mathcal{A}$, and let $\mathcal{B}$ be a maximal abelian self-adjoint subalgebra of $\mathcal{M}'$. If $\mathcal{N}$ is a von Neumann algebra with $\mathcal{M} \subset \mathcal{N} \subset \mathcal{B}'$, then the embedding $I_\mathcal{B}$ of $\mathcal{L}^n(\mathcal{M},\mathcal{N}\colon/\mathcal{A})$ into*

$$\mathcal{L}^n(C^*(\mathcal{M},\mathcal{B}), \mathcal{B}'\colon/C^*(\mathcal{A},\mathcal{B}))$$

defined in Lemma 5.2.2 induces a map $I_\mathcal{B}$ from $H^n(\mathcal{M},\mathcal{N}\colon/\mathcal{A})$ into

$$H^n(C^*(\mathcal{M},\mathcal{B}), \mathcal{B}'\colon/C^*(\mathcal{A},\mathcal{B})) = 0.$$

Proof. By Lemma 5.2.2, $I_\mathcal{B}$ and ∂ commute. Thus $I_\mathcal{B}$ induces a map from $H^n(\mathcal{M},\mathcal{N}\colon/\mathcal{A})$ into

$$H^n(C^*(\mathcal{M},\mathcal{B}), \mathcal{B}'\colon/C^*(\mathcal{A},\mathcal{B})),$$

taking $\mathcal{N}$ for the module $\mathcal{V}$ and noting that $(\mathcal{B}\mathcal{N})^- \subseteq \mathcal{B}'$. Now observe that $\mathcal{B}'$ is a dual normal $\mathcal{B}'$-module, so

$$H^n(C^*(\mathcal{M},\mathcal{B}), \mathcal{B}'\colon/C^*(\mathcal{A},\mathcal{B})) \cong H^n(\mathcal{B}',\mathcal{B}'\colon/C^*(\mathcal{A},\mathcal{B}))$$
$$\cong H_w^n(\mathcal{B}',\mathcal{B}'\colon/C^*(\mathcal{A},\mathcal{B}))$$

by applying Theorem 3.3.1 to the C^*-algebra $C^*(\mathcal{M},\mathcal{B})$ with weak closure $\mathcal{B}'$, and then to $\mathcal{B}'$ itself. Now the weak continuity of the cocycles in H_w^n implies that

$$H_w^n(\mathcal{B}',\mathcal{B}'\colon/C^*(\mathcal{A},\mathcal{B})) \cong H_w^n(\mathcal{B}',\mathcal{B}'\colon/\mathcal{B}')$$

which is clearly zero (see the proof of Corollary 3.4.6).

5.2.4 Corollary. Let M be a von Neumann algebra with centre Z and von Neumann subalgebra N containing Z, and let B be a masa in M'. If ϕ is in $\mathcal{L}_w^n(M, M: /N)$ with $\partial\phi = 0$, then there is a ψ in

$$\mathcal{L}_w^{n-1}(C^*(M, B), B': /C^*(N, B))$$

such that $I_B\phi = \partial\psi$.

Proof. Regarding ϕ as a continuous, rather than a normal, linear map, shows that $I_B\phi = \partial\chi$ for some

$$\chi \in \mathcal{L}^{n-1}(C^*(M, B), B': /C^*(A, B))$$

by Proposition 5.2.3. Now by Theorem 3.3.1 the natural embedding induces an isomorphism

$$H_w^n(C^*(M, B), B': /C^*(A, B)) \text{ onto } H_c^n(C^*(M, B), B': /C^*(A, B)).$$

Hence there is a map $\psi \in \mathcal{L}_w^{n-1}(C^*(M, B), B': /C^*(N, B))$ such that $I_B\phi = \partial\psi$.

Compare the following corollary, where there is the restriction $N \subset B'$, with Theorem 4.2.6, where the cohomology groups are assumed to be completely bounded.

5.2.5 Corollary. Let M be a von Neumann algebra and let B be a maximal abelian self-adjoint subalgebra in M'. If N is a hyperfinite von Neumann algebra such that $M \subseteq N \subseteq B'$, then

$$H^n(M, N) = H_w^n(M, N) = 0.$$

Proof. If $\phi \in \mathcal{L}^n(M, N)$ with $\partial\phi = 0$, then by Proposition 5.2.3 there exists $\psi \in \mathcal{L}^{n-1}(M, B')$ such that $\phi = \partial\psi$. Let E be a conditional expectation from $B(H)$ onto N. Then E is an N-module map. Hence $E\partial = \partial E$ so that

$$\phi = E\phi = \partial E\psi,$$

and $E\psi \in \mathcal{L}^{n-1}(M, N)$ as required.

5.3 Popa Subfactors and Masas

5.3.1 Remarks. This section contains three results due to Popa concerning injective subfactors and masas of type II_1 factors. The results hold for type II_1 von Neumann algebras with separable predual but are proved here for factors. In Theorem 5.3.7 the hyperfinite subfactor N in the type II_1 factor

$\mathcal{M}$ and a masa $\mathcal{A}$ in $\mathcal{M}$ are constructed together so that $\mathcal{N}' \cap \mathcal{M} = \mathbb{C}1$. However there is no control over the structure of $\mathcal{A}$, which is constructed as a limit of finite dimensional subalgebras of a changing sequence of masas in $\mathcal{M}$. The second main theorem (5.3.10) gives the existence of $\mathcal{N}$ containing a Cartan subalgebra; one has to ensure that the Cartan normalization property holds in the construction of $\mathcal{N}$. The third theorem (5.3.12) is that the abelian von Neumann algebra $\mathcal{B} = (\mathcal{A} \cup J\mathcal{A}J)''$ is a masa on $L^2(\mathcal{M})$ is $\mathcal{A}$ is a Cartan algebra in $\mathcal{M}$.

Throughout this section $\mathcal{M}$ will denote a type II_1 factor with separable predual and trace tr with $tr(1) = 1$. Let

$$\|x\|_2 = tr(x^*x)^{1/2} \quad (x \in \mathcal{M})$$

denote the Hilbert space norm induced by tr on $\mathcal{M}$, and let $L^2(\mathcal{M})$ denote the completion of $\mathcal{M}$ in this $\|\cdot\|_2$-norm. If $\mathcal{N}$ is a von Neumann subalgebra of $\mathcal{M}$, let $E_{\mathcal{N}}$ denote the trace preserving conditional expectation from $\mathcal{M}$ onto $\mathcal{N}$. Note that at the $L^2(\mathcal{M})$ level, $E_{\mathcal{N}}$ is just the restriction to $\mathcal{M}$ of the orthogonal projection $e_{\mathcal{N}}$ from $L^2(\mathcal{M})$ onto the closure of $\mathcal{N}$ in $L^2(\mathcal{M})$. Hence

$$\|x - E_{\mathcal{N}}(x)\|_2 = \inf\{\|x - y\|_2\colon y \in \mathcal{N}\}$$

for all x in $\mathcal{M}$ by the corresponding property for the orthogonal projection on $L^2(\mathcal{M})$.

5.3.2 Lemma. *Let $(\mathcal{A}_n)$ be an increasing sequence of finite dimensional abelian $*$-subalgebras of $\mathcal{M}$ and let $\mathcal{A}$ be the weak closure of $\cup \mathcal{A}_n$.*

(1) Then $\mathcal{A}$ is maximal abelian in $\mathcal{M}$ if and only if

$$\|E_{\mathcal{A}'_n \cap \mathcal{M}}(x) - E_{\mathcal{A}_n}(x)\|_2 \to 0 \text{ as } n \to \infty \text{ for all } x \text{ in } \mathcal{M}.$$

(2) If $\mathcal{A}$ is maximal abelian, then

$$\|E_{\mathcal{A}'_n \cap \mathcal{M}}(x) - E_{\mathcal{A}}(x)\|_2 \to 0 \text{ as } n \to \infty \text{ for all } x \text{ in } \mathcal{M}.$$

Proof. We shall show that

(a) $\qquad \|E_{\mathcal{A}_n}(x) - E_{\mathcal{A}}(x)\|_2 \to 0 \text{ as } n \to \infty \quad (x \in \mathcal{M}), \quad$ and

(b) $\qquad \|E_{\mathcal{A}'_n \cap \mathcal{M}}(x) - E_{\mathcal{A}' \cap \mathcal{M}}(x)\|_2 \to 0 \text{ as } n \to \infty \quad (x \in \mathcal{M}).$

Note that $\mathcal{A}$ is maximal abelian in $\mathcal{M}$ if and only if $E_{\mathcal{A}} = E_{\mathcal{A}' \cap \mathcal{M}}$, which links (a), (b) and the conclusion of the lemma.

(a) The sequence $(\mathcal{A}_n)$ is an increasing sequence of finite dimensional subspaces in $L^2(\mathcal{M})$ with the $\|\cdot\|_2$-closure of $\cup\mathcal{A}_n$ equal to the $\|\cdot\|_2$-closure of $\mathcal{A}$ since $\mathcal{A}$ is strongly closed in $\mathcal{M}$ and the map $x \to \|x\|_2$ is strongly continuous on bounded sequences. At the $L^2(\mathcal{M})$ level, the projections $E_{\mathcal{A}_n}$ converge strongly to $E_\mathcal{A}$ so (a) follows. This uses the observation that $E_\mathcal{N}$ is the restriction of the Hilbert space $L^2(\mathcal{N})$ projection to $\mathcal{M}$ and the following elementary property of Hilbert space. If $Y_1 \subset Y_2 \subset \cdots$ are closed subspaces of a Hilbert space H with Y_0 equal to the closure of $\cup Y_n$, then

$$\|(e_n - e_0)\eta\| \to 0 \text{ as } n \to \infty \quad (\eta \in H)$$

where e_n is the orthogonal projection onto Y_n. This comes from the little calculation that for all $\eta \in H$

$$\inf\{\|e_0\eta - \nu\|: \ \nu \in \cup Y_n\} = \inf_n \inf\{\|e_0\eta - \nu\|: \ \nu \in Y_n\}$$
$$= \inf_n \|e_0\eta - e_n\eta\|$$
$$= \lim_n \|e_0\eta - e_n\eta\|$$

since $(\|e_0\eta - e_n\eta\|)$ is a decreasing sequence.

(b) Observe that

$$\bigcap_n (\mathcal{A}'_n \cap \mathcal{M}) = \left(\bigcap_n \mathcal{A}'_n\right) \cap \mathcal{M} = (\cup\mathcal{A}_n)' \cap \mathcal{M} = \mathcal{A}' \cap \mathcal{M}.$$

Note that $(E_{\mathcal{A}'_n \cap \mathcal{M}}(x))$ is a bounded sequence in $\mathcal{M}$ in the (uniform) norm $\|\cdot\|$ so has a subnet converging weakly to some y in $\mathcal{M}$. This y is in $\mathcal{A}' \cap \mathcal{M}$. For each n, $E_{\mathcal{A}'_n \cap \mathcal{M}}(y) = y$ so

$$\|E_{\mathcal{A}'_n \cap \mathcal{M}}(x) - y\|_2^2 = tr(E_{\mathcal{A}'_n \cap \mathcal{M}}((x - y)^*) \cdot (E_{\mathcal{A}'_n \cap \mathcal{M}}(x - y)))$$
$$= tr((x^* - y^*) \cdot (E_{\mathcal{A}'_n \cap \mathcal{M}}(x) - y)).$$

Letting n run over the net for which $E_{\mathcal{A}'_n \cap \mathcal{M}}(x)$ converges weakly to y implies that the last term in the equation tends to zero over the net. Hence

$$\|E_{\mathcal{A}'_n \cap \mathcal{M}}(x) - E_{\mathcal{A}' \cap \mathcal{M}}(x)\|_2 \leq \|E_{\mathcal{A}'_n \cap \mathcal{M}}(x) - y\|_2$$

tends to zero over the net; this inequality used the orthogonality of the projection $E_{\mathcal{A}' \cap \mathcal{M}}$ in the $\|\cdot\|_2$-norm. Part (2) of the conclusion follows directly from (b) above.

The following lemma splits off a part of the proof of Lemma 5.3.4 as a standard separate calculation.

5.3.3 Lemma. *Let A be a masa in a type II_1 factor M with separable predual. If F is a maximal totally ordered set of projections in A, then $\{tr(f): f \in F\} = [0,1]$.*

Proof. Let λ be in the closure of $\{tr(f): f \in F\}$. Then there is an increasing or decreasing sequence $tr(f_n)$, with $f_n \in F$, converging to λ. Since F is totally ordered, the sequence (f_n) is an increasing or decreasing sequence of projections so converges strongly to a projection $e \in A$. Suppose f_n is an increasing sequence; the decreasing case is similar. If g is in F, then either $f_n g = f_n$ for all n or there is an N such that $f_n g = g$ for all $n \geq N$; in the former case $eg = e$, in the latter $eg = g$. Thus the projection e is comparable with all projections in F so $F \cup \{e\}$ is a totally ordered family of projections. By maximality, $e \in F$. Hence $\{tr(f): f \in F\}$ is closed since $tr(e) = \lambda$ by the strong continuity of the trace. If the set $\{tr(f): f \in F\}$ were disconnected by omitting the open interval $(tr(g_1), tr(g_2))$ with $g_1 \leq g_2$ in F, then there is a non-zero projection h in A with $h \leq g_2 - g_1$ and

$$0 < tr(h) < tr(g_2 - g_1)$$

since A has no minimal projections. Hence

$$g_1 \leq g_1 + h \leq g_2$$

and

$$tr(g_1) < tr(g_1 + h) < tr(g_2),$$

so $F \cup \{g_1 + h\}$ is a totally ordered family of projections in A properly containing F. The resulting contradiction proves the lemma.

5.3.4 Lemma. *If A is a masa in a type II_1 factor M with separable predual, then there exists a $*$-isomorphism θ from A onto $L^\infty[0,1]$ such that*

$$tr(x) = \int_0^1 \theta(x)(t)dt$$

for all $x \in A$.

Proof. By the separability of the predual of M, and hence of A, there is a sequence $P = \{p_n: n = 0, 1, \ldots\}$ of projections in A such that the algebra B_0 generated by this set of projections is weakly dense in A. Assume that $p_0 = 1$. A totally ordered set G of projections in A will be constructed from P by induction so that the algebras generated by G and P are equal and G is the image of the dyadic rationals in [0,1] under an order preserving mapping $t \mapsto q(t)$. Let $q(0) = 0$, $q(1) = 1$ and $q(1/2) = p_1$. Then

$$q(0) \leq q(1/2) \leq q(1)$$

and
$$\text{Alg}\{q(j \cdot 2^{-1}): \ j = 0, 1, 2\} = \text{Alg}\{p_0, p_1\}.$$

Suppose that the map q from $\{j2^{-n}: \ 0 \le j \le 2^n\}$ into $\mathcal{A}_0$ has been defined such that $0 \le j \le k \le 2^n$ implies that

$$q(j \cdot 2^{-n}) \le q(k \cdot 2^{-n})$$

and
$$\text{Alg}\{q(j \cdot 2^{-n}): \ 0 \le j \le 2^n\} = \text{Alg}\{p_i: \ 0 \le i \le n\}.$$

The next step in the induction requires the definition of $q((2j + 1)2^{-n-1})$ which is done by splitting the difference between the projections $q((j + 1)2^{-n})$ and $q(j \cdot 2^{-n})$ using p_{n+1}. Let

$$
(1) \quad
\begin{aligned}
q((2j + 1)2^{-n-1}) &= q(j \cdot 2^{-n} + 2^{-n-1}) \\
&= q(j \cdot 2^{-n}) + p_{n+1}\{q((j + 1)2^{-n}) - q(j2^{-n})\}
\end{aligned}
$$

for $0 \le j \le n - 1$.

Then for these values of j,

$$q(j \cdot 2^{-n}) \le q((2j + 1) \cdot 2^{-n-1}) \le q((j + 1)2^{-n})$$

so the map q preserves the order. Note that the inductive definition implies directly that

$$(2) \qquad \text{Alg}\{q(j \cdot 2^{-n-1}): \ 0 \le j \le 2^{n+1}\} \subseteq \text{Alg}\{p_i: \ 0 \le i \le n + 1\}.$$

Observe that the equation

$$
\begin{aligned}
p_{n+1} &= \sum_{j=0}^{2^n - 1} (q((j + 1)2^{-n}) - q(j \cdot 2^{-n}))p_{n+1} \\
&= \sum_{j=0}^{2^n - 1} (q((2j + 1) \cdot 2^{-n-1}) - q(j \cdot 2^{-n}))
\end{aligned}
$$

holds by (1) so that

$$p_{n+1} \in \text{Alg}\{q(j \cdot 2^{-n-1}): \ 0 \le j \le 2^{n+1}\}$$

and the two algebras in (2) are equal.

By Zorn's Lemma, choose a maximal totally ordered set $\mathcal{F}$ of projections in $\mathcal{A}$ with $\mathcal{G} \subseteq \mathcal{F}$. Then $\text{Alg}(\mathcal{F}) \supseteq \text{Alg}(\mathcal{G}) = \text{Alg}(\mathcal{P})$ so that $\text{Alg}(\mathcal{F})$ is weakly dense in $\mathcal{A}$. By Lemma 5.3.3 the trace is an injective order preserving map from $\mathcal{F}$ onto $[0,1]$. This is the stage at which we move from $\mathcal{G}$ with its unusual dyadic order to the usual order of $\mathcal{A}$. The $*$-isomorphism θ is defined from $\mathrm{Alg}(\mathcal{F})$ into $L^\infty[0,1]$ by

$$\theta(f) = \chi_{[0,tr(f)]},$$

where χ_W is the characteristic function of a subset W of $[0,1]$. Note that θ is a $*$-isomorphism because if $f_1, f_2 \in \mathcal{F}$, then

$$tr(f_1 f_2) = \min(tr(f_1), tr(f_2))$$

which matches the product of the characteristic functions. Clearly $\|\theta x\|_\infty = \|x\|$ for all x in $\mathrm{Alg}(\mathcal{F})$ because the norm in $\mathcal{A}_0$ is a C^*-algebra norm. The equation

$$\int_0^1 \theta(f)(t)dt = tr(f)$$

for all f in $\mathcal{F}$ extends by linearity to $\mathrm{Alg}(\mathcal{F})$ so

$$\int_0^1 \theta(x)(t)dt = tr(x)$$

for all x in $\mathrm{Alg}(\mathcal{F})$. Thus $\|\theta(x)\|_2 = \|x\|_2$ for all x in $\mathrm{Alg}(\mathcal{F})$. Since the $\|\cdot\|$-unit ball of $\mathrm{Alg}(\mathcal{P}) \subseteq \mathrm{Alg}(\mathcal{F})$ is dense in the $\|\cdot\|$-unit ball B of $\mathcal{A}$ in the $\|\cdot\|_2$-norm by the Kaplansky density theorem and by the equivalence of the $\|\cdot\|_2$-norm and strong topologies on B, it follows that θ extends to a $*$-homomorphism θ from $\mathcal{A}$ into $L^\infty[0,1]$ with

$$\int_0^1 \theta(x)(t)dt = tr(x)$$

for all x in $\mathcal{A}$. The map θ is surjective because $\theta(\mathrm{Alg}\ \mathcal{F})$ is strongly dense in $L^\infty[0,1]$. This proves the lemma.

5.3.5 Corollary. *Let $\mathcal{A}_0$ be a finite dimensional subalgebra of a masa $\mathcal{A}$ in a type II_1 factor M such that all of the minimal projections in $\mathcal{A}_0$ are equivalent (in M). Then $\mathcal{A}$ is the strong closure of the union of an increasing sequence of finite dimensional subalgebras $\mathcal{A}_n$ of $\mathcal{A}$ with $\mathcal{A}_0 \subseteq \mathcal{A}_1$, and each $\mathcal{A}_n$ having the property that all its minimal projections are equivalent.*

Proof. Two projections in M are equivalent if and only if they have the same trace. The corollary now follows directly from Lemma 5.3.4 and the corresponding property for $L^\infty[0,1]$; in $L^\infty[0,1]$ the projections (characteristic functions) are assumed to have the same integral.

5.3.6 Definition. A masa $\mathcal{A}$ in a type II_1 von Neumann algebra M is called a *Cartan subalgebra* of M if the unitary normalizer

$$\mathcal{N}(\mathcal{A}) = \{u \in \mathcal{U}(M)\colon uAu^* = \mathcal{A}\}$$

of $\mathcal{A}$ in M generates M as a von Neumann algebra.

5.3.7 Theorem. *Let $\mathcal{M}$ be a type II_1 factor with separable predual. There is a masa $\mathcal{A}$ in $\mathcal{M}$ and a hyperfinite subfactor $\mathcal{N}$ of $\mathcal{M}$ containing $\mathcal{A}$ such that $\mathcal{A}$ is a Cartan subalgebra of $\mathcal{N}$ and $\mathcal{N}' \cap \mathcal{M}$ is $\mathbb{C}1$.*

Proof. Let $\{x_j : j \in \mathbb{N} \cup \{0\}\}$ be a dense sequence in the closed unit ball of $\mathcal{M}$ in the $\|\cdot\|_2$-norm. Such a sequence exists by the separability of the predual of $\mathcal{M}$ and the equivalence of the $\|\cdot\|_2$-topology and strong topology on this closed unit ball. Assume $x_0 = 1$. By induction on n, increasing sequences of

(1) finite dimensional abelian subalgebras $A_0 = \mathbb{C}1 \subset A_1 \subset A_2 \subset \cdots$ and

(2) matrix subalgebras
$$\mathcal{N}_0 = \mathbb{C}1 \subset \mathcal{N}_1 \subset \cdots$$

of $\mathcal{M}$ are constructed with matrix units e_{ij}^n $(1 \leq i, j \leq k_n)$ in $\mathcal{N}_n$ such that

(3) $\{e_{ii}^n : 1 \leq i \leq k_n\}$ is the set of minimal projections in A_n, i.e., A_n is the diagonal of $\mathcal{N}_n$ with respect to this basis,

(4) for each $1 \leq s \leq n - 1$, the natural unitary in $\mathcal{N}_s$ that interchanges e_{11}^s and e_{jj}^s $(1 \leq j \leq k_s)$ normalizes A_n, and

(5) $\|E_{A_n' \cap \mathcal{M}}(x_j) - E_{A_n}(x_j)\|_2 \leq 2^{-n}$ for $1 \leq j \leq n$.

Suppose that A_n, $\mathcal{N}_n$, e_{ij}^n $(1 \leq i, j \leq k_n)$ have been constructed. Let $\mathcal{B}$ be a masa in $\mathcal{M}$ containing A_n. By Corollary 5.3.5, applied to A_n in place of $\mathcal{A}$, there is an increasing sequence $\mathcal{B}_\ell$ of finite dimensional subalgebras of $\mathcal{B}$ containing A_n with the properties of that lemma. As $\mathcal{B}$ is a masa in $\mathcal{M}$ with $\mathcal{B}$ equal to the weak closure of $\cup \mathcal{B}_\ell$, Lemma 5.3.2 implies that

$$\|E_{\mathcal{B}' \cap \mathcal{M}}(x) - E_{\mathcal{B}_\ell}(x)\|_2 \to 0 \text{ as } \ell \to \infty$$

for all $x \in \mathcal{M}$. Choose ℓ such that

(6) $$\|(E_{\mathcal{B}' \cap \mathcal{M}} - E_{\mathcal{B}_\ell})(e_{ij}^n x_t e_{j1}^n)\|_2 < 2^{-(n+1)} k_n^{-1/2}$$

for $1 \leq t \leq n + 1$ and $1 \leq j \leq k_n$. Let e_{ii}^{n+1} $(1 \leq i \leq k_{n+1})$ denote the set of minimal projections in $\mathcal{B}_\ell$. Since these projections are all equivalent,

$$tr(e_{ii}^{n+1}) = tr(e_{11}^{n+1})$$

for all i. Further, the minimal projections e_{ii}^n $(1 \leq i \leq k_n)$ satisfy

$$tr(e_{ii}^n) = tr(e_{11}^n)$$

and are contained in $\mathcal{B}_\ell$. Hence k_n divides k_{n+1} and each e_{ii}^n is the sum of $k_{n+1}/k_n = m_n$ minimal projections e_{jj}^{n+1}. By renumbering if necessary, we can ensure that $e_{11}^n = \sum_1^{m_n} e_{jj}^{n+1}$. The projections e_{jj}^{n+1} $(1 \leq j \leq m_n)$ are all equivalent so there are partial isometrics v_j with

$$v_j^* v_j = e_{11}^{n+1} \text{ and } v_j v_j^* = e_{jj}^{n+1} \text{ for } 1 \leq j \leq m_n.$$

Let $e_{ij}^{n+1} = v_i v_j^*$ for $1 \leq i, j \leq m_n$. Then e_{ij}^{n+1} $(1 \leq i, j \leq m_n)$ form a set of matrix units.

Let $\mathcal{A}_{n+1}$ be the linear subspace of $\mathcal{M}$ spanned by the set

$$\{e_{j1}^n e_{tt}^{n+1} e_{1j}^n \colon 1 \leq t \leq m_n, \ 1 \leq j \leq k_n\},$$

and let $\mathcal{N}_{n+1}$ be the linear subspace spanned by the set

$$\{e_{i1}^n e_{st}^{n+1} e_{1j}^n \colon 1 \leq s, t \leq m_n, \ 1 \leq i, j \leq k_n\}.$$

Then $\mathcal{A}_{n+1}$ is an abelian subalgebra of $\mathcal{M}$ whose minimal projections are the elements of its spanning set, $\mathcal{A}_n \subseteq \mathcal{A}_{n+1}$, $\mathcal{A}_{n+1} \subseteq \mathcal{N}_{n+1}$, and the minimal projections in $\mathcal{A}_{n+1}$ are all equivalent. Further, $\mathcal{N}_{n+1}$ is a subalgebra of $\mathcal{M}$ naturally isomorphic to $\mathbf{M}_{k_{n+1}}$, where $k_{n+1} = m_n k_n$, with matrix units $e_{i1}^n e_{st}^{n+1} e_{1j}^n$ for $1 \leq s, t \leq m_n$ and $1 \leq i, j \leq k_n$.

Let $1 \leq s \leq n-1$, let $1 \leq j \leq k_s$ and let u be the natural unitary in $\mathcal{N}_s$ that interchanges e_{11}^s and e_{jj}^s. Then u normalizes $\mathcal{A}_n$ by (4) so normalizes the minimal projections $e_{i1}^n e_{tt}^{n+1} e_{1i}^n$ $(1 \leq i \leq k_n, \ 1 \leq t \leq m_n)$ spanning $\mathcal{A}_{n+1}$. If u is the natural unitary in $\mathcal{N}_n$ that interchanges e_{11}^n and e_{ii}^n, then u normalizes $\mathcal{A}_{n+1}$ by definition of the minimal projections in $\mathcal{A}_{n+1}$.

Now it is necessary to check property (5). The projections e_{jj}^n are in the algebras $\mathcal{A}_{n+1}' \cap \mathcal{M}$ and $\mathcal{A}_{n+1}$ so

$$(E_{\mathcal{A}_{n+1}' \cap \mathcal{M}} - E_{\mathcal{A}_{n+1}})(x) = \sum e_{jj}^n (E_{\mathcal{A}_{n+1}' \cap \mathcal{M}} - E_{\mathcal{A}_{n+1}})(x) e_{jj}^n$$

for all x in $\mathcal{M}$. The pairwise orthogonality of the subalgebras $e_{jj}^n \mathcal{M} e_{jj}^n$ of $\mathcal{M}$ implies that

$$\|(E_{\mathcal{A}_{n+1}' \cap \mathcal{M}} - E_{\mathcal{A}_{n+1}})(x)\|_2^2 = \sum_{j=1}^{k_n} \|e_{jj}^n (E_{\mathcal{A}_{n+1}' \cap \mathcal{M}} - E_{\mathcal{A}_{n+1}})(x) e_{jj}^n\|_2^2.$$

The module property of the conditional expectations implies that the last sum equals

$$\sum_{j=1}^{k_n} \|(E_{\mathcal{A}_{n+1}' \cap \mathcal{M}} - E_{\mathcal{A}_{n+1}})(e_{jj}^n x e_{jj}^n)\|^2.$$

The map $e_{jj}^n x e_{jj}^n \to e_{1j}^n x e_{j1}^n$ is a $\|\cdot\|_2$-isometric isomorphism from $e_{jj}^n \mathcal{M} e_{jj}^n$ onto $e_{1j}^n \mathcal{M} e_{j1}^n$ that carries

$$e_{jj}^n(\mathcal{A}_{n+1}' \cap \mathcal{M}) e_{jj}^n \text{ onto } e_{1j}^n(\mathcal{A}_{n+1}' \cap \mathcal{M}) e_{j1}^n$$

and $e_{jj}^n \mathcal{A}_{n+1} e_{jj}^n$ onto $e_{1j}^n \mathcal{A}_{n+1} e_{j1}^n$. Thus

$$\|(E_{\mathcal{A}_{n+1}' \cap \mathcal{M}} - E_{\mathcal{A}_{n+1}})(e_{jj}^n x e_{jj}^n)\|_2 = \|(E_{\mathcal{A}_{n+1}' \cap \mathcal{M}} - E_{\mathcal{A}_{n+1}})(e_{1j}^n x e_{j1}^n)\|_2.$$

These equations now imply that

$$\|(E_{\mathcal{A}_{n+1}' \cap \mathcal{M}} - E_{\mathcal{A}_{n+1}})(x)\|_2^2 = \sum \|(E_{\mathcal{A}_{n+1}' \cap \mathcal{M}} - E_{\mathcal{A}_{n+1}})(e_{1j}^n x e_{j1}^n)\|_2^2.$$

The definition of $\mathcal{A}_{n+1}$ in terms of $\mathcal{B}_\ell$ and its minimal projections implies that

$$e_{11}^n(\mathcal{A}_{n+1}' \cap \mathcal{M}) e_{11}^n = e_{11}^n(\mathcal{B}_\ell \cap \mathcal{M}) e_{11}^n$$

and

$$e_{11}^n \mathcal{A}_{n+1} e_{11}^n = e_{11}^n \mathcal{B}_\ell e_{11}^n.$$

Thus

$$\|(E_{\mathcal{A}_{n+1}' \cap \mathcal{M}} - E_{\mathcal{A}_{n+1}})(x)\|^2 = \sum \|(E_{\mathcal{B}_\ell' \cap \mathcal{M}} - E_{\mathcal{B}_\ell})(e_{1j}^n x e_{j1}^n)\|_2^2.$$

Replacing x by x_t for $1 \le t \le n+1$ and using (6) leads to

$$\|(E_{\mathcal{A}_{n+1}' \cap \mathcal{M}} - E_{\mathcal{A}_{n+1}})(x_t)\|_2^2 \le \sum 4^{-(n+1)} \cdot k_n^{-1} = 4^{-(n+1)},$$

which proves (5).

Let $\mathcal{N}$ be the weak closure of $\cup \mathcal{N}_n$ and let $\mathcal{A}$ be the weak closure of $\cup \mathcal{A}_n$. Clearly $\mathcal{N}$ is a hyperfinite factor, $\mathcal{A}$ is abelian and $\mathcal{A} \subseteq \mathcal{N}$. By (5) and the $\|\cdot\|_2$-density of $\{x_j : j \ge 0\}$ in the closed unit ball of $\mathcal{M}$, it follows that $\|(E_{\mathcal{A}_n' \cap \mathcal{M}} - E_{\mathcal{A}_n})(x)\|_2$ tends to zero as n tends to infinity for all x in $\mathcal{M}$. Hence $\mathcal{A}$ is a masa in $\mathcal{M}$ by Lemma 5.3.2. Now

$$\mathcal{N}' \cap \mathcal{M} \subseteq \mathcal{A}' \cap \mathcal{M} = \mathcal{A}$$

since $\mathcal{A}$ is a masa in $\mathcal{M}$ so

$$\mathcal{N}' \cap \mathcal{M} \subseteq \mathcal{N}' \cap \mathcal{A} \subseteq \mathcal{N}' \cap \mathcal{N} = \mathbb{C}1.$$

Finally note that if $n \in \mathbb{N}$ and $1 \le j \le k_n$, then the natural unitary u in $\mathcal{N}_n$ that interchanges e_{11}^n and e_{jj}^n normalizes $\mathcal{A}$ since it normalizes $\mathcal{A}_s$ for all $s \ge n+1$ (by (4)). As j varies over $1 \le j \le k_n$ these unitaries generate $\mathcal{N}_n$.

Hence the unitary normalizer of $\mathcal{A}$ in $\mathcal{N}$ generates $\mathcal{N}$ as a von Neumann algebra. The proof is complete.

5.3.8 Notation and Remarks. For the remainder of this section $\mathcal{A}$ denotes a Cartan subalgebra of a type II_1 factor $\mathcal{M}$. Recall that a Cartan subalgebra $\mathcal{A}$ of $\mathcal{M}$ is a masa in $\mathcal{M}$ such that the unitary normalizer $\mathcal{N}(\mathcal{A}) = \{u \in \mathcal{U}(\mathcal{M}): u\mathcal{A}u^* = \mathcal{A}\}$ of $\mathcal{A}$ in $\mathcal{M}$ generates $\mathcal{M}$ as a von Neumann algebra. Let

$$\mathcal{G} = \{v \in \mathcal{M}: v \text{ is a partial isometry, } v\mathcal{A}v^* \subseteq \mathcal{A} \text{ and } v^*\mathcal{A}v \subseteq \mathcal{A}\}.$$

Let v, w be in $\mathcal{G}$. Clearly

$$vw\mathcal{A}(vw)^*\mathcal{A} \subseteq \mathcal{A} \quad \text{and} \quad (vw)^*\mathcal{A}vw \subseteq \mathcal{A}.$$

Since $\mathcal{A}$ is commutative and v^*v and ww^* are in $\mathcal{A}$, we have

$$v^*vww^* = ww^*v^*v.$$

The following calculation shows that vww^*v^* is a projection:

$$\begin{aligned}
(vww^*v^*)(vww^*v^*) &= v(ww^*)(v^*v)ww^*v^* \\
&= v(v^*v)ww^*ww^*v^* \\
&= vww^*v^*.
\end{aligned}$$

Thus vw is a partial isometry. If $\mathcal{P}(\mathcal{A})$ denotes the set of projections in $\mathcal{A}$, observe that

$$\{up: u \in \mathcal{N}(\mathcal{A}),\ p \in \mathcal{P}(\mathcal{A})\}$$

is contained in $\mathcal{G}$ because $up\mathcal{A}pu^* \subseteq \mathcal{A}$ and $pu^*\mathcal{A}up \subseteq \mathcal{A}$. Note that $\mathcal{G} = \mathcal{G}^*$. Let w_1, w_2 be elements of $\mathcal{G}$ which satisfy

$$w_1w_1^* \cdot w_2w_2^* = 0, \quad w_1^*w_1 \cdot w_2^*w_2 = 0,$$

and define w to be $w_1 + w_2$. Then w is a partial isometry and

$$\begin{aligned}
w\mathcal{A}w^* &= w_1\mathcal{A}w_1^* + w_2\mathcal{A}w_2^* + w_1w_1^*w_1\mathcal{A}w_2^*w_2w_2 \\
&\quad + w_2w_2^*w_2\mathcal{A}w_1^*w_1w_1^* \\
&= w_1\mathcal{A}w_1^* + w_2\mathcal{A}w_2^* \\
&\subseteq \mathcal{A}
\end{aligned}$$

because the other two terms are zero. By the symmetry w to w^* it follows that $w^*\mathcal{A}w \subseteq \mathcal{A}$.

The following lemma ensures that the hyperfinite algebra $\mathcal{N}$ constructed in Theorem 5.3.10 is generated by the unitary normalizer of $\mathcal{A}$ in it.

5.3.9 Lemma. If $\{f_1, \ldots, f_k\}$ is a set of orthogonal equivalent projections in a Cartan subalgebra $\mathcal{A}$ of $\mathcal{M}$, then there are matrix units f_{ij} $(1 \leq i, j \leq k)$ in $\mathcal{G}$ such that $f_{ii} = f_i$ for $1 \leq i \leq k$.

Proof. The first step of the proof is to show that if e_1 and e_2 are equivalent orthogonal projections in $\mathcal{A}$, then there is an element $v \in \mathcal{G}$ with $v^*v = e_1$ and $vv^* = e_2$. The proof of this is a standard maximality technique employed in von Neumann algebras. For example, it is similar to the construction of the central support of a projection.

Let
$$\mathcal{F} = \{w \in \mathcal{G}: w^*w \leq e_1 \quad \text{and} \quad ww^* \leq e_2\}.$$

Note that $0 \in \mathcal{F}$ so $\mathcal{F}$ is not empty. Define a partial order $\preceq$ on $\mathcal{F}$ by

$$w_1 \preceq w_2 \text{ if and only if } w_1^*w_1 \leq w_2^*w_2 \text{ and } w_1 = w_2 w_1^* w_1.$$

Observe that this does define a partial order on $\mathcal{F}$ with the transitivity following directly from the definition of the partial order. By Zorn's Lemma there is a maximal totally ordered subfamily $\mathcal{F}_0$ of $\mathcal{F}$. Note that $\{w^*w: w \in \mathcal{F}_0\}$ forms an increasing net of projections in $\mathcal{A}$ that converge strongly to p_0 in $\mathcal{P}(\mathcal{A})$. Further, the net $w \in \mathcal{F}_0$ converges weakly to a partial isometry w_0 in $\mathcal{M}$ with $w_0^*w_0 = p_0 \leq e_1$ and $w_0w_0^* \leq e_2$. Then $w_0\mathcal{A}w_0^* \subseteq \mathcal{A}$ and $w_0^*\mathcal{A}w_0 \subseteq \mathcal{A}$ so that $w_0 \in \mathcal{F}$ by the definition of $\mathcal{F}$, and $w \preceq w_0$ for all $w \in \mathcal{F}_0$ so $w_0 \in \mathcal{F}_0$ is maximal in $\mathcal{F}$. Suppose that $w_0^*w_0 \neq e_1$. Let $p_1 = e_1 - w_0^*w_0$ and $p_2 = e_2 - w_0w_0^*$. Note that p_1 and p_2 are equivalent orthogonal projections; the orthogonality comes from that of e_1 and e_2, and the equivalence from $tr(p_1) = tr(p_2)$.

If $p_1 u p_2 u^* = 0$ for all unitaries $u \in \mathcal{N}(\mathcal{A})$, then the non-zero projection

$$g = \vee\{up_2u^*: u \in \mathcal{N}(\mathcal{A})\}$$

in $\mathcal{A}$ satisfies $p_1 g = 0$ and $p_2 g = p_2$. Further g is in $\mathcal{M} \cap \mathcal{M}'$ since $\mathcal{N}(\mathcal{A})$ generates $\mathcal{M}$ as a von Neumann algebra. Hence $g = 1$, which implies that $p_1 = 0$ contrary to $tr(p_1) \neq 0$. This contradiction shows that there is an element u in $\mathcal{N}(\mathcal{A})$ such that $p_1 u p_2 u^* = p_3$ is not zero. Now $p_3 \leq p_1$ and $u^*p_3 \in \mathcal{N}(\mathcal{A}) \cdot \mathcal{P}(\mathcal{A}) \subseteq \mathcal{G}$ by Remark 5.3.8. Also

$$u^*p_3u \leq p_2 = e_2 - w_0w_0^*$$

so

$$w_0w_0^* \cdot u^*p_3 \cdot p_3u = 0,$$

and

$$p_3 \leq p_1 = e_1 - w_0^*w_0$$

so $w_0^* w_0 \cdot p_3 u u^* p_3 = 0$. Hence the partial isometry $w_0 + u^* p_3$ is in $\mathcal{G}$ by Remark 5.3.8 and is strictly greater than w_0. This contradicts the maximality of w_0 so implies that $w_0^* w_0 = e_1$. Thus $w_0 w_0^* = e_2$ since

$$tr(e_2 - w_0 w_0^*) = tr(e_1) - tr(w_0^* w_0) = 0.$$

Using the above property of $\mathcal{G}$, choose f_{1j} $(1 \leq j \leq k)$ in $\mathcal{G}$ such that

$$f_{1j} f_{1j}^* = f_1 \quad \text{and} \quad f_{1j}^* f_{1j} = f_j \text{ for } 1 \leq j \leq k.$$

Define $f_{ij} = f_{1i}^* f_{1j}$ for $1 \leq i, j \leq k$. These elements are the required matrix units in $\mathcal{G}$.

5.3.10 Theorem. *Let M be a type II_1 factor with separable predual and let A be a Cartan subalgebra of M. Then there is a hyperfinite subfactor N of M with $N' \cap M = \mathbb{C}1$ such that A is a Cartan subalgebra of N.*

Proof. The idea of the proof is to construct an increasing sequence of matrix algebras using Lemmas 5.3.2 and 5.3.9 to ensure that the weak closure of the union has the required properties.

Let $\{a_n : n \in \mathbb{N}\}$ be a sequence of projections in A whose linear span is dense in A in the strong operator topology. Note that A contains such a sequence by Lemma 5.3.3. By induction, we construct an increasing sequence

$$N_0 = \mathbb{C}1 \subseteq N_1 \subseteq N_2 \ldots$$

of finite dimensional matrix subalgebras of M with matrix units (e_{ij}^n) satisfying

(1) $e_{ij}^n \in \mathcal{G}$ for $1 \leq i, j \leq k(n)$, where N_n is isomorphic to $M_{k(n)}(\mathbb{C})$,

(2) if A_n is the finite dimensional subalgebra generated by the set

$$\{e_{ii}^n : 1 \leq i \leq k(n)\},$$

then $A_n \subseteq A$ and

(3) $$\|a_j - E_{A_n}(a_j)\|_2 \leq 2^{-n} \text{ for } 1 \leq j \leq n.$$

As usual E_{A_n} denotes the trace preserving conditional expectation from M onto A_n.

Suppose that N_n has been constructed. Since $w \in \mathcal{G}$ implies that $wAw^* \subseteq A$ by construction of $\mathcal{G}$ in (5.3.8), the finite set

$$\mathcal{F} = \{e_{1i}^n a_t e_{i1}^n : 1 \leq t \leq n+1, \ 1 \leq i \leq k(n)\}$$

is contained in $\mathcal{A}$. By Corollary 5.3.5, there is a finite dimensional subalgebra $\mathcal{B}$ of $\mathcal{A}$ that contains $\mathcal{A}_n$ such that

$$(4) \qquad \|(I - E_{\mathcal{B}})(e_{1i}^n a_j e_{i1}^n)\|_2 \leq k(n) 2^{-(n+1)}$$

for $1 \leq i \leq k(n)$ and $1 \leq j \leq n+1$, and

(5) the minimal projections $f_1, \ldots, f_m$ in $\mathcal{B}$ are all equivalent.

The projection e_{11}^n is in $\mathcal{A}_n \subseteq \mathcal{B}$, so is a sum of some of the minimal projections in $\mathcal{B}$, and e_{11}^n acts as an identity for the finite set

$$\{e_{1i}^n a_t e_{i1}^n \colon 1 \leq i \leq k(n), \ 1 \leq t \leq n+1\}.$$

Thus those f_j that do not occur in the sum for e_{11}^n may be discarded, and we assume that

$$(6) \qquad e_{11}^n = f_1 + \cdots + f_m.$$

By Lemma 5.3.9 there are matrix units f_{ij}, $1 \leq i, j \leq m$, such that $f_{ii} = f_i$ for $1 \leq i \leq m$. The matrix units $e_{k\ell}^{n+1}$ ($1 \leq k, \ell \leq mk(n)$) are defined by using a tensor product idea. Let

$$(7) \qquad e_{i,j,s,t}^{n+1} = e_{s1}^n f_{ij} e_{1t}^n \text{ for } 1 \leq s, t \leq k(n) \text{ and } 1 \leq i, j \leq m,$$

and let $\mathcal{N}_{n+1}$ be the linear span of these matrix units. Then $\mathcal{N}_{n+1}$ is isomorphic to the algebra of $k(n+1) \times k(n+1)$ matrices, where $k(n+1) = mk(n)$ and $\mathcal{N}_{n+1}$ contains $\mathcal{N}_n$. The diagonal of $\mathcal{N}_{n+1}$ is contained in $\mathcal{A}$ since each e_{ij}^n is in $\mathcal{G}$.

To finish the induction we need to prove inequality (3) from (4). If L_x and R_y denote respectively left and right multiplication by x, observe that $L_{e_{i1}^n} R_{e_{1i}^n}$ is a $\| \cdot \|_2$-norm isometry from $e_{1i}^n \mathcal{A} e_{i1}^n$ onto $e_{ii}^n \mathcal{A} e_{ii}^n$. Using the $\mathcal{A}_{n+1}$-module property of $E_{\mathcal{A}_{n+1}}$ and the equation

$$e_{11}^n \mathcal{A}_{n+1} e_{11}^n = e_{11}^n \mathcal{B} e_{11}^n,$$

it follows that

$$(8) \qquad \|(I - E_{\mathcal{A}_{n+1}})e_{ii}^n x e_{ii}^x\|_2 = \|(I - E_{\mathcal{A}_{n+1}})e_{1i}^n x e_{1i}^n\|_2 = \|(I - E_{\mathcal{B}})e_{1i}^n x e_{1i}^n\|_2$$

for all x in $\mathcal{A}$. Since $\mathcal{A}$ is abelian and e_{ii}^n ($1 \leq i \leq k(n)$) are orthogonal in $\mathcal{A}_{n+1}$,

$$
\begin{aligned}
(9) \qquad \|x - E_{\mathcal{A}_{n+1}}(x)\|_2^2 &= \left\| \sum e_{ii}^n (I - E_{\mathcal{A}_{n+1}})(x) e_{ii}^n \right\|_2^2 \\
&= \sum \|e_{ii}^n (I - E_{\mathcal{A}_{n+1}})(x) e_{ii}^n\|_2^2 \\
&= \sum \|(I - E_{\mathcal{A}_{n+1}})(e_{ii}^n x e_{ii}^n)\|_2^2 \\
&= \sum \|(I - E_{\mathcal{B}})(e_{1i}^n x e_{i1}^n)\|_2^2
\end{aligned}
$$

for all x in $\mathcal{A}$, by (8), where the sums are taken over $1 \leq i \leq k(n)$. Inequalities (4) and (9) give (3) directly; this was the reason for the choice of ε in (4).

Let $\mathcal{N}$ be the weak closure of $\cup \mathcal{N}_n$; then $\mathcal{N}$ is a subfactor of $\mathcal{M}$. Note that the weak closure of $\cup \mathcal{A}_n$ is equal to $\mathcal{A}$ by inequality (3) and the choice of $\{a_n : n \in \mathsf{N}\}$. Let u_{ij}^n be the unitary in $\mathcal{N}_n$ that interchanges e_{ii}^n and e_{jj}^n and leaves the remaining minimal projections unchanged. Then $u_{ij}^n \mathcal{A} u_{ij}^{n*} = \mathcal{A}$ because the parts $e_{kk}^n \mathcal{A}$, $k \neq i, j$, are untouched by the unitary and the interchange of e_{ii}^n and e_{jj}^n carried out by u_{ij}^n is just that of e_{ij}^n, which is in $\mathcal{G}$ by (1). Hence the unitary normalizer $\mathcal{N}(\mathcal{A})$ of $\mathcal{A}$ in $\mathcal{N}$ contains the set

$$\{u_{ij}^n : n \in \mathsf{N}, 1 \leq i, j \leq k(n)\},$$

which generates $\mathcal{N}$ as a von Neumann algebra. Thus $\mathcal{A}$ is a Cartan subalgebra of $\mathcal{N}$. Finally

$$\mathcal{N}' \cap \mathcal{M} \subseteq \mathcal{A}' \cap \mathcal{M} = \mathcal{A} \subseteq \mathcal{N}$$

since $\mathcal{A}$ is a masa in $\mathcal{M}$. Hence

$$\mathcal{N}' \cap \mathcal{M} \subseteq \mathcal{N}' \cap \mathcal{N} = \mathbb{C}1$$

which finishes the proof.

The final theorem (5.3.12) of this section gives a masa in $B(H)$ from a Cartan subalgebra of $\mathcal{M}$. As before, $\mathcal{M}$ is a type II_1 von Neumann factor with separable predual and faithful normal trace tr, and $H = L^2(\mathcal{M}, tr)$ is the completion of $\mathcal{M}$ in the $\|\cdot\|_2$-norm. Let J be the usual involution on H defined by $Jx = x^*$ for all x in $\mathcal{M}$. If V is a linear subspace of H (usually of $\mathcal{M}$), then p_V denotes the orthogonal projection from H onto the $\|\cdot\|_2$-norm closure V^- of V. If E is the trace preserving conditional expectation from $\mathcal{M}$ onto a von Neumann subalgebra $\mathcal{C}$, then $Ex = p_{\mathcal{C}}x$ for all x in $\mathcal{M}$. Recall that

$$\mathcal{N}(\mathcal{A}) = \{u : u \text{ unitary in } \mathcal{M}, \ u\mathcal{A}u^* = \mathcal{A}\}.$$

5.3.11 Lemma. *Let $\mathcal{A}$ be a masa in a type II_1 factor $\mathcal{M}$ with separable predual and faithful normal trace tr. If $\mathcal{B}$ is the von Neumann algebra generated by $\mathcal{A}$ and $J\mathcal{A}J$ and if $u \in \mathcal{N}(\mathcal{A})$, then*

(1) $$\mathcal{A}u = u\mathcal{A} = \mathcal{A}u\mathcal{A} = \mathcal{A} \cdot J\mathcal{A}J \cdot u$$

(2) $$p_{\mathcal{A}u} = p_{\mathcal{B}u} = p_{\mathcal{B}'u} \in \mathcal{B} \quad \text{and}$$

(3) $$\mathcal{A}p_{\mathcal{A}u} = \mathcal{B}p_{\mathcal{A}u} = \mathcal{B}'p_{\mathcal{A}u}.$$

Proof. Since u is in $\mathcal{N}(\mathcal{A})$,

$$u\mathcal{A} = u\mathcal{A}(u^*u) = (u\mathcal{A}u^*)u = \mathcal{A}u$$

so

$$\mathcal{A} \cdot J\mathcal{A}J \cdot u = \mathcal{A}u\mathcal{A} = \mathcal{A}\mathcal{A}u = \mathcal{A}u,$$

(as subspaces of $L^2(\mathcal{M}, tr)$), which proves (1).

We shall now prove (2) and (3) when $u = 1$, regarded as an element of H. By Lemma 5.3.3 and the structure of $L^\infty[0,1]$, there is a double sequence $\{e_j^n \colon 1 \le j \le 2^n, \, n \ge 0\}$ of projections in $\mathcal{A}$ such that

1) $e_1^0 = 1$, $e_j^n = e_{2j-1}^{n+1} + e_{2j}^{n+1}$ for $1 \le j \le 2^n$ and $n \ge 0$,

2) the linear span of $\{e_j^n \colon 1 \le j \le 2^n, n \ge 0\}$ is weakly dense in $\mathcal{A}$. For each $n \ge 0$, let

3) $q_n = \sum_{j=1}^{2^n} e_j^n J e_j^n J$ be a projection in $\mathcal{B}$.

If x is in $\mathcal{M}$, then

$$q_n x = \sum_j e_j^n x e_j^n = E_{\mathcal{A}_n' \cap \mathcal{M}}(x),$$

where $\mathcal{A}_n$ is the linear span of $\{e_j^n \colon 1 \le j \le 2^n\}$ and $E_{\mathcal{A}}x = p_{\mathcal{A}}x$. Since $\mathcal{A}$ is a masa in $\mathcal{M}$ and the union of the $\mathcal{A}_n$ is weakly dense in $\mathcal{A}$,

$$\|q_n x - p_{\mathcal{A}}x\|_2 = \|E_{\mathcal{A}_n' \cap \mathcal{M}}x - E_{\mathcal{A}}x\|_2 \to 0$$

as $n \to \infty$ for all x in $\mathcal{M}$ by Lemma 5.3.2.

Since $\|q_n\| = 1$ for all n, it follows that q_n converges strongly to $p_{\mathcal{A}}$ as n tends to infinity so that $p_{\mathcal{A}}$ is in $\mathcal{B}$. By (1),

$$\mathcal{A}^- = (\mathcal{A}1)^- = (\mathcal{B}1)^-$$

so $p_{\mathcal{A}1} = p_{\mathcal{B}1}$. Since $p_{\mathcal{B}'1}$ is the smallest projection p in $\mathcal{B}$ with $p1 = 1$ and since $p_{\mathcal{A}}1 = 1$, we have $p_{\mathcal{B}'1} \le p_{\mathcal{A}1}$. From the inclusions

$$\mathcal{B}'1 \supseteq \mathcal{B}1 \supseteq \mathcal{A}1 = \mathcal{A}$$

it follows that

$$p_{\mathcal{B}'1} \ge p_{\mathcal{B}1} \ge p_{\mathcal{A}1}.$$

Thus

$$p_{\mathcal{B}'1} = p_{\mathcal{B}1} = p_{\mathcal{A}1}$$

proving (2) for $u = 1$.

Since the closure of $\mathcal{A}$ in $L^2(\mathcal{M}, tr)$ is

$$p_{\mathcal{A}1}H = p_{\mathcal{A}1}L^2(\mathcal{M}, tr) = L^2(\mathcal{A}, tr|_{\mathcal{A}}),$$

the restriction of $\mathcal{A}$ to $p_{\mathcal{A}1}H$ is the standard form of $\mathcal{A}$. Hence $\mathcal{A}$ restricted to $p_{\mathcal{A}1}H$ is a masa in $B(p_{\mathcal{A}1}H)$. Also $\mathcal{B}p_{\mathcal{A}1} = \mathcal{B}p_{\mathcal{B}1}$ is an abelian algebra on $p_{\mathcal{A}1}H$ containing the algebra $\mathcal{A}p_{\mathcal{A}1}$ so $\mathcal{B}p_{\mathcal{A}1} = \mathcal{A}p_{\mathcal{A}1}$. Since $\mathcal{B}p_{\mathcal{A}1}$ is a masa in $B(p_{\mathcal{A}1}H)$ and $\mathcal{B}'p_{\mathcal{A}1} = \mathcal{B}'p_{\mathcal{B}'1}$ is in the commutant of $\mathcal{B}p_{\mathcal{A}1}$ in $B(p_{\mathcal{A}1}H)$, it follows that $\mathcal{B}'p_{\mathcal{A}1} = \mathcal{B}p_{\mathcal{A}1}$. This proves (3) for $u = 1$.

To deduce the general case of $u \in \mathcal{N}(\mathcal{A})$ from the one for $u = 1$ we use the following elementary Hilbert space fact. If p is the (orthogonal) projection from H onto a closed subspace K and u is a unitary operator on H, then upu^* is the projection from H onto uK. If u is in $\mathcal{N}(\mathcal{A})$, define the unitary u on H by $x \mapsto ux$ for all $x \in \mathcal{M}$. The inner automorphism $b \mapsto ubu^*$ on $B(H)$ leaves $\mathcal{A}$ invariant, is the identity automorphism on $J\mathcal{A}J$ and so leaves $\mathcal{B}$ invariant. Further $up_{\mathcal{A}}u^* = p_{u\mathcal{A}} = p_{\mathcal{A}u}$ and similarly for $p_{\mathcal{B}1}$ and $p_{\mathcal{B}'1}$. Equations (2) and (3) now follow for u from the corresponding ones for 1.

5.3.12 Theorem. *If $\mathcal{A}$ is a Cartan subalgebra in a type II_1 factor $\mathcal{M}$ with separable predual, then $\mathcal{B} = (\mathcal{A} \cup J\mathcal{A}J)''$ is a masa in $B(L^2(\mathcal{M}))$.*

Proof. Since $\mathcal{A}$ is a Cartan subalgebra of $\mathcal{M}$, $\mathcal{M}$ is generated as a von Neumann algebra by $\mathcal{N}(\mathcal{A})$ and hence the closed linear span of $\{\mathcal{A}u \colon u \in \mathcal{N}(\mathcal{A})\}$ is dense in $L^2(\mathcal{M})$ in $\|\cdot\|_2$. Thus the smallest projection greater than all the projections $p_{\mathcal{A}u}$ ($u \in \mathcal{N}(\mathcal{A})$) is the identity operator. By Lemma 5.3.11, $p_{\mathcal{A}u} \in \mathcal{B} \subseteq \mathcal{B}'$ and $\mathcal{B}p_{\mathcal{A}u} = \mathcal{B}'p_{\mathcal{A}u}$ for all $u \in \mathcal{N}(\mathcal{A})$ so that $\mathcal{B} = \mathcal{B}'$. Hence $\mathcal{B}$ is a masa in $B(L^2(\mathcal{M}))$ since $\mathcal{B}$ is abelian.

5.4 The Pisier–Haagerup–Grothendieck Inequality for Module Maps

The non-commutative Grothendieck inequality due to Pisier [Pi1] and Haagerup [Ha3] is one of the fundamental geometrical results about C^*-algebras. We shall assume the basic form of this theorem below and use it to prove various strengthened versions under additional assumptions on the bilinear forms and operators. Under suitable hypotheses this leads to the complete boundedness of various bilinear maps (see Theorem 1.6.2). Though the hypotheses of these results may seem rather restrictive they will be achieved by various natural operators in cohomological calculations.

There are detailed proofs of the non-commutative Grothendieck inequality in Pisier's book [Pi2], [Ha3] and [KS] though the best constants are only obtained in [Ha3] and [Pi2].

5.4.1 Theorem. *Let* $\mathcal{A}$ *and* $\mathcal{B}$ *be* C^*-*algebras and let* ϕ *be a continuous bilinear form on* $\mathcal{A} \times \mathcal{B}$. *Then there exist states* f_1, f_2 *on* $\mathcal{A}$ *and* g_1, g_2 *on* $\mathcal{B}$ *such that*

$$|\phi(x,y)| \le \|\phi\|(f_1(x^*x) + f_2(xx^*))^{\frac{1}{2}} \cdot (g_1(y^*y) + g_2(yy^*))^{\frac{1}{2}}.$$

for all $x \in \mathcal{A}$ *and* $y \in \mathcal{B}$.

For von Neumann algebras the forms are usually required to be normal as these respect the ultraweak topology on the algebra. Using the normal plus singular decomposition of states, [T, p. 127], Haagerup [Ha3] obtained the following corollary, which is the crucial first step in the subsequent cohomology calculations. We give this proof.

5.4.2 Corollary. *Let* $\mathcal{M}$ *and* $\mathcal{N}$ *be von Neumann algebras and let* ϕ *be a normal bilinear form from* $\mathcal{M} \times \mathcal{N}$ *into* $\mathbb{C}$. *Then there exist normal states* f_1, f_2 *on* $\mathcal{M}$ *and* g_1, g_2 *on* $\mathcal{N}$ *such that*

$$|\phi(x,y)| \le \|\phi\|(f_1(x^*x) + f_2(xx^*))^{\frac{1}{2}}(g_1(y^*y) + g_2(yy^*))^{\frac{1}{2}}$$

for all $x \in \mathcal{M}$ *and* $y \in \mathcal{N}$.

Proof. Using a standard von Neumann maximality argument we first show that if f is a singular positive linear functional on a von Neumann algebra $\mathcal{M}$, then there is an increasing net (p_α) of projections in $\mathcal{M}$ such that p_α tends to 1 strongly and $f(p_\alpha) = 0$ for all α.

Let (e_i) be a maximal set of non-zero pairwise orthogonal projections in $\mathcal{M}$ such that $f(e_i) = 0$ for all i; Zorn's Lemma gives such a family. A standard result [T, p. 134] on singular positive linear functionals is that if e is a projection in $\mathcal{M}$, then there is a non-zero projection q in $\mathcal{M}$ with $q \le e$ and $f(q) = 0$. If $1 - \sum e_i \ne 0$, then there is a non-zero projection q in $\mathcal{M}$ with $f(q) = 0$ and $q \le 1 - \sum e_i$; this projection q may be adjoined to the family (e_i) contradicting the maximality of this family. Thus $\sum e_i = 1$. The family (p_α) is now obtained by taking all finite sums of the projections e_i from the maximal family chosen.

By the Pisier–Haagerup non-commutative Grothendieck Theorem (5.4.1) there are states f_1, f_2 on $\mathcal{M}$ and g_1, g_2 on $\mathcal{N}$ such that

$$(1) \qquad |\phi(x,y)| \le \|\phi\|(f_1(x^*x) + f_2(xx^*))^{\frac{1}{2}}(g_1(y^*y) + g_2(y^*y))^{\frac{1}{2}}$$

for all $x, y \in \mathcal{M}$. Let $f_j = f_{jn} + f_{js}$ be the splitting of f_j into its normal (f_{jn}) and singular (f_{js}) parts [T, p. 127]. The positive functional $f = f_{1s} + f_{2s}$ is also singular. Now choose a net (p_α) of increasing projections in $\mathcal{M}$ such that p_α tends to 1 strongly and $f(p_\alpha) = 0$ for all α. Thus

$$f_{1s}(p_\alpha) = f_{2s}(p_\alpha) = 0$$

for all α. The positivity of f_{js} implies that

$$f_{js}(p_\alpha x p_\alpha) = 0$$

for all α and all $x \in \mathcal{M}$. The product on $\mathcal{M}$ is strongly continuous on bounded sets and normal functionals are strongly continuous. Hence,

$$
\begin{aligned}
\text{(2)} \qquad & \lim_\alpha (f_1((p_\alpha x p_\alpha)^* p_\alpha x p_\alpha) + f_2(p_\alpha x p_\alpha (p_\alpha x p_\alpha)^*)) \\
= \ & \lim_\alpha (f_{1n}(p_\alpha x^* p_\alpha x p_\alpha) + f_2(p_\alpha x p_\alpha x^* p_\alpha)) \\
= \ & f_{1n}(x^* x) + f_{2n}(x x^*)
\end{aligned}
$$

for each $x \in \mathcal{M}$. Since ϕ is a normal functional in the first variable with the second one fixed,

$$|\phi(x, y)| = \lim_\alpha |\phi(p_\alpha x p_\alpha, y)|,$$

so that by (1) and (2) above

$$|\phi(x, y)| \le \|\phi\|(f_{1n}(x^* x) + f_{2n}(x x^*))^{\frac{1}{2}} \cdot (g_1(y^* y) + g_2(y y^*))^{\frac{1}{2}}$$

for all x and y in $\mathcal{M}$.

The same proof, applied to this inequality with $\mathcal{M}$ and $\mathcal{N}$ interchanged, leads to

$$|\phi(x, y)| \le \|\phi\|(f_{1n}(x^* x) + f_{2n}(x x^*))^{\frac{1}{2}} (g_{1n}(y^* y) + g_{2n}(y y^*))^{\frac{1}{2}}$$

where g_{jn} is the normal part of g_j $(j = 1, 2)$. Now

$$\|f_{nj}\| \le \|f_j\| = 1 \text{ and } \|g_{jn}\| \le \|g_j\| = 1,$$

so if f_{jn} and g_{jn} are not states then multiplying by the inverse of their norms gives the required normal states f_j' and g_j'. This proves the normal bilinear form of the Pisier–Haagerup inequality.

There are several routes to proving the following result, which is required subsequently, but given (5.4.2), the following is the shortest proof. This proof corresponds exactly to the deduction of the "little Grothendieck inequality" concerned with operators from $C(\Omega)$ into a Hilbert space from the Grothendieck inequality for bilinear forms on $C(\Omega)$.

5.4.3 Corollary. *If ψ is an ultraweakly–weakly continuous linear operator from a von Neumann algebra $\mathcal{M}$ into a Hilbert space H, then there are normal states f and g on $\mathcal{M}$ such that*

$$\|\psi(x)\|^2 \le \|\psi\|(f(x x^*) + g(x^* x))$$

for all $x \in \mathcal{M}$.

Proof. The bilinear form ϕ defined on $\mathcal{M} \times \mathcal{M}$ by

$$\phi(x, y) = \langle \psi(x), \psi(y^*) \rangle$$

is separately ultraweakly continuous with $\|\phi\| = \|\psi\|^2$. Hence there are normal states f_1, f_2, g_1, g_2 on $\mathcal{M}$ such that

$$|\phi(x, y)| \leq \|\phi\| (f_1(x^*x) + f_2(xx^*))^{\frac{1}{2}} \cdot (g_1(y^*y) + g_2(yy^*))^{\frac{1}{2}}.$$

Using the inequality $(\alpha\beta)^{1/2} \leq (\alpha + \beta)/2$ between the geometric mean and the arithmetic mean leads to

$$\begin{aligned} \|\psi(x)\|^2 &= |\phi(x, x^*)| \\ &\leq 1/2 \|\psi\|^2 (f_1(x^*x) + f_2(xx^*) + g_1(xx^*) + g_2(x^*x)) \end{aligned}$$

for all $x \in \mathcal{M}$. Letting

$$f = f_1 + g_2 \quad \text{and} \quad g = f_2 + g_1$$

completes the proof. Note that the switch in the subscripts is forced by the $*$ on the second variable in the definition of ϕ, which is there to ensure the bilinearity of ϕ.

The above two corollaries are now used in the next few results for type II_1 von Neumann algebras that are modules over large injective subalgebras. The idea is to remove the linear functional in Corollaries 5.4.2 and 5.4.3 that is associated with the anti-representation by averaging over the injective subalgebra, so introducing a trace. The tracial functional enables us to switch the order of x and x^*, thus reducing to representations, the Haagerup tensor product and eventually complete boundedness. Firstly we require a convenient description of the normal centre-valued trace or centre-valued tracial conditional expectation in terms of a hyperfinite subfactor with trivial relative commutant.

If $\mathcal{M}$ is a type II_1 (or finite) von Neumann algebra with centre $\mathcal{Z}$, let τ denote the centre-valued tracial conditional expectation constructed in such a situation [T], [Di2], [StrZ], [KR4]. Then τ is a normal conditional expectation from $\mathcal{M}$ onto $\mathcal{Z}$ with the tracial property

$$\tau(xy) = \tau(yx) \quad \text{for all } x, y \in \mathcal{M}.$$

Further each tracial $(f(xy) = f(yx))$ normal linear functional f on $\mathcal{M}$ is of the form

$$f(x) = g(\tau(x))$$

for some normal linear functional g on $\mathcal{Z}$.

5.4.4 Lemma. *Let $\mathcal{M}$ be a type II_1 von Neumann algebra containing a hyperfinite von Neumann algebra $\mathcal{N}$, whose relative commutant $\mathcal{N}' \cap \mathcal{M}$ is the centre $\mathcal{Z}$ of $\mathcal{M}$, and let τ be the normal centre-valued tracial conditional expectation on $\mathcal{M}$. Let $\{\mathcal{N}_\lambda : \lambda \in \Lambda\}$ be an increasing net of finite dimensional $*$-subalgebras of $\mathcal{N}$, whose union is ultraweakly dense in $\mathcal{M}$. Let $\mathcal{U}_\lambda$ denote the unitary group of $\mathcal{N}_\lambda$ and let μ_λ denote normalized Haar measure on $\mathcal{U}_\lambda$. For each $x \in \mathcal{M}$,*

$$\tau(x) = \lim_{\lambda \in \Lambda} \int_{\mathcal{U}_\lambda} uxu^* d\mu(u)$$

ultraweakly.

Proof. Suppose that there exists an element x in $\mathcal{M}$ for which either equality does not hold above or the limit does not exist. Let

$$x_\lambda = \int_{\mathcal{U}_\lambda} uxu^* d\mu_\lambda(u)$$

for each $\lambda \in \Lambda$. Note that the net $\{x_\lambda : \lambda \in \Lambda\}$ of elements in $\mathcal{M}$ is uniformly bounded by $\|x\|$ by the normalization of the Haar measures μ_λ. Since $\tau(x)$ is not the ultraweak limit of the net $\{x_\lambda : \lambda \in \Lambda\}$ there is a normal state $f \in \mathcal{M}_*$, an $\varepsilon > 0$ and a cofinal subnet Δ of Λ such that

(1) $$|f(\tau(x) - x_\delta)| \geq \varepsilon$$

for all $\delta \in \Delta$. Closed balls in $\mathcal{M}$ are ultraweakly compact so there exists a cofinal subnet Γ of Δ such that the net $\{x_\gamma : \gamma \in \Gamma\}$ converges ultraweakly to an element y in $\mathcal{M}$. Taking ultraweak limits in (1) over $\gamma \in \Gamma$ yields

(2) $$|f(\tau(x) - y)| \geq \varepsilon.$$

Now normality and the tracial property of τ imply that

(3) $$\begin{aligned}
\tau(x_\gamma) &= \tau\left(\int_{\mathcal{U}_\gamma} uxu^* d\mu_\gamma(u) \right) \\
&= \int_{\mathcal{U}_\gamma} \tau(uxu^*) d\mu_\gamma(u) \\
&= \int_{\mathcal{U}_\gamma} \tau(x) d\mu_\gamma(u) \\
&= \tau(x)
\end{aligned}$$

for all $\gamma \in \Gamma$. Taking the limit over Γ in (3) gives

(4) $$\tau(y) = \tau(x).$$

The invariance of Haar measure shows that $x_\gamma \in \mathcal{N}'_\gamma$, since it is just the image of a conditional expectation onto $\mathcal{N}'_\gamma$. Thus $x_\gamma \in \mathcal{N}'_\lambda$ for $\lambda \leq \gamma$ since the nesting $\mathcal{N}_\lambda \subseteq \mathcal{N}_\gamma$ gives $\mathcal{N}'_\lambda \supseteq \mathcal{N}'_\gamma$. Hence $y \in \mathcal{N}'_\lambda$ for each $\lambda \in \Lambda$ since Γ is a cofinal subnet of Λ. Now $(\cup\{\mathcal{N}_\lambda : \lambda \in \Lambda\})' = \mathcal{N}'$ so $y \in \mathcal{N}' \cap \mathcal{M} = \mathcal{Z}$. Hence $\tau(y) = y$, which gives a contradiction with (2) and (4).

The following lemma ensures that suitable normal bilinear module maps from $\mathcal{M} \times \mathcal{M}$ into $B(H)$ lift to the Haagerup tensor product $\mathcal{M} \otimes_h \mathcal{M}$ into $B(H)$.

5.4.5 Lemma. *Let $\mathcal{M}$ be a type II_1 von Neumann algebra containing a hyperfinite von Neumann subalgebra $\mathcal{N}$ whose relative commutant $\mathcal{N}' \cap \mathcal{M}$ is the centre $\mathcal{Z}$ of $\mathcal{M}$.*

(i) If ϕ is a normal bilinear form on $\mathcal{M} \times \mathcal{M}$ satisfying

$$\phi(xa, y) = \phi(x, ay)$$

for all $x, y \in \mathcal{M}$ and $a \in \mathcal{N}$, then there are normal states f and g on $\mathcal{M}$ such that
$$|\phi(x,y)| \leq 2\|\phi\| f(xx^*)^{\frac{1}{2}} g(y^*y)^{\frac{1}{2}}.$$

(ii) If ψ is a normal bilinear map from $\mathcal{M} \times \mathcal{M}$ into $B(H)$, for some Hilbert space H, satisfying
$$\psi(xa, y) = \psi(x, ay)$$

for all $x, y \in \mathcal{M}$ and $a \in \mathcal{N}$, then

$$\left\| \sum_1^n \psi(x_j, y_j) \right\| \leq 2\|\psi\| \left\| \sum_1^n x_j x_j^* \right\|^{\frac{1}{2}} \left\| \sum_1^n y_j^* y_j \right\|^{\frac{1}{2}}$$

for all $x_1, \ldots, x_n, y_1, \ldots, y_n \in \mathcal{M}$.

Proof. (i) By the normal bilinear form version of the non-commutative Grothendieck inequality (5.4.2) there are normal states f_1, f_2, g_1, g_2 on $\mathcal{M}$ such that

$$|\phi(x,y)| \leq \|\phi\|(f_1(x^*x) + f_2(xx^*))^{\frac{1}{2}} \cdot (g_1(y^*y) + g_2(yy^*))^{\frac{1}{2}}$$

for all $x, y \in \mathcal{M}$. Let $\{\mathcal{N}_\lambda : \lambda \in \Lambda\}$ be an increasing net of finite dimensional *-subalgebras whose union is ultraweakly dense in $\mathcal{N}$. By Lemma 5.4.4,

(1) $$\tau(x) = \lim_{\lambda \in \Lambda} \int_{\mathcal{U}_\lambda} uxu^* d\mu_\lambda(u)$$

ultraweakly for all $x \in \mathcal{M}$, where τ is the centre-valued trace on $\mathcal{M}$.

If u is in $\mathcal{U}_\lambda$, then

$$|\phi(x,y)| = |\phi(xu^*, uy)|$$
$$\leq \|\phi\|(f_1(ux^*xu^*) + f_2(xx^*))^{\frac{1}{2}} \cdot (g_1(y^*y) + g_2(uyy^*u^*))^{\frac{1}{2}}$$
$$\leq \|\phi\| \cdot 1/2[f_1(ux^*xu^*) + f_2(xx^*) + g_1(y^*y) + g_2(uyy^*u^*)]$$

using the domination of the geometric mean by the arithmetic mean. Now integrating this over $\mathcal{U}_\lambda$ and using the continuity of f_1 and g_2 gives

$$|\phi(x,y)| \leq \|\phi\| \cdot 1/2 \left[f_1 \left(\int_{\mathcal{U}_\lambda} ux^*xu^* d\mu_\lambda(u) \right) \right.$$
$$\left. + f_2(xx^*) + g_1(y^*y) + g_2 \left(\int_{\mathcal{U}_\lambda} uyy^*u^* d\mu_\lambda(u) \right) \right].$$

Taking the limit over λ and using the normality of f_1 and g_2 implies that

$$|\phi(x,y)| \leq \|\phi\| \cdot 1/2[f_1(\tau(x^*x)) + f_2(xx^*) + g_1(y^*y) + g_2(\tau(yy^*))]$$
$$= \|\phi\| \cdot 1/2[(f_1\tau + f_2)(xx^*) + (g_1 + g_2\tau)(y^*y)]$$

by the tracial property of τ. Let $f = \frac{1}{2}(f_1\tau + f_2)$ and $g = \frac{1}{2}(g_1 + g_2\tau)$ so

$$|\phi(x,y)| \leq \|\phi\|(f(xx^*) + g(y^*y))$$

for all $x, y \in \mathcal{M}$. A standard quadratic argument with x replaced by $t^{1/2}x$ and y by $t^{-1/2}y$ for all $t > 0$ leads to

$$|\phi(x,y)| \leq 2\|\phi\| f(xx^*)^{\frac{1}{2}} g(y^*y)^{\frac{1}{2}}$$

as required

(ii) Let ξ and η be unit vectors in H and let

$$\phi(x,y) = \langle \psi(x,y)\xi, \eta \rangle$$

for all $x, y \in \mathcal{M}$. Then ϕ is a normal bilinear form on $\mathcal{M}$ satisfying the hypotheses of (i). Hence there are normal states f and g on $\mathcal{M}$ such that

$$|\phi(x,y)| \leq 2\|\phi\| f(xx^*)^{\frac{1}{2}} g(y^*y)^{\frac{1}{2}}$$

for all $x, y \in \mathcal{M}$. It follows that

$$\left| \left\langle \sum_1^n \psi(x_j, y_j)\xi, \eta \right\rangle \right| \leq \sum_1^n |\phi(x_j, y_j)|$$

$$\leq 2\|\phi\| \sum_1^n f(x_j x_j^*)^{\frac{1}{2}} g(y_j^* y_j)^{\frac{1}{2}}$$

$$\leq 2\|\psi\| \left(f\left(\sum_1^n x_j x_j^* \right) \right)^{\frac{1}{2}} \left(g\left(\sum_1^n y_j^* y_j \right) \right)^{\frac{1}{2}}$$

$$\leq 2\|\psi\| \left\| \sum_1^n x_j x_j^* \right\|^{\frac{1}{2}} \left\| \sum_1^n y_j^* y_j \right\|^{\frac{1}{2}},$$

using the Cauchy–Schwarz inequality.

Now taking the supremum over all unit vectors ξ and η in H proves the inequality in (ii).

Observe that a standard sesquilinear form argument based on ϕ may be used to show that the ϕ in Lemma 5.4.5(i) has a representation of the following type. There are normal representations π and θ of $\mathcal{M}$ on Hilbert spaces H_π and H_θ, unit vectors $\xi \in H_\pi$ and $\eta \in H_\theta$ and a continuous linear operator $T\colon H_\theta \to H_\pi$ with $\|T\| \leq 2\|\phi\|$ such that

$$\phi(x, y) = \langle \pi(x) T \theta(y) \xi, \eta \rangle$$

for all $x, y \in \mathcal{M}$ and

$$\pi(a)T = T\theta(a)$$

for all $a \in \mathcal{N}$.

Lemma 5.4.5 will be used to obtain two related but different results (5.4.6 and 5.4.7) that enable us to show that suitable maps are completely bounded. The complete boundedness will ensure that certain cohomology groups are zero by Theorem 4.3.1.

5.4.6 Lemma. *Let $\mathcal{M}$ be a type II_1 von Neumann algebra with centre $\mathcal{Z}$, with Cartan subalgebra $\mathcal{A}$ and with a hyperfinite subalgebra $\mathcal{N}$ containing $\mathcal{A}$ as a Cartan subalgebra such that $\mathcal{N}' \cap \mathcal{M} = \mathcal{Z}$. Moreover, assume that $\mathcal{M}$ has a separable predual. Let τ be a faithful normal trace on $\mathcal{M}$ and regard $\mathcal{M}$ as acting on the separable Hilbert space $H = L^2(\mathcal{M}, \tau)$. Let J be the involution on $L^2(\mathcal{M}, \tau)$ defined by $Jx = x^*$ ($x \in \mathcal{M}$). Let $\mathcal{B}$ denote the algebra $J\mathcal{A}J$. If ψ is a normal bilinear $\mathcal{N}$-module map from $\mathcal{M} \times \mathcal{M}$ into $\mathcal{B}' = (J\mathcal{A}J)'$, then ψ is completely bounded and $\|\psi\|_{cb} \leq 2\|\psi\|$.*

Proof. By Lemma 5.2.2 there is a continuous bilinear map $\psi_1 = I_{\mathcal{B}}\psi$ from $C^*(\mathcal{M}, \mathcal{B}) \times C^*(\mathcal{M}, \mathcal{B})$ into $B(H)$ satisfying

$$\psi_1(m_1 b_1, m_2 b_2) = \psi(m_1, m_2) b_1 b_2$$

for all $m_1, m_2 \in \mathcal{M}$ and all $b_1, b_2 \in \mathcal{B}$. Further $\|\psi_1\| = \|\psi\|$. By Lemma 5.4.5 the bilinear map ψ on $\mathcal{M} \times \mathcal{M}$ satisfies

$$(1) \qquad \left\| \sum_1^n \psi(x_j, y_j) \right\| \leq 2\|\psi\| \cdot \left\| \sum_1^n x_j x_j^* \right\|^{\frac{1}{2}} \left\| \sum_1^n y_j^* y_j \right\|^{\frac{1}{2}}$$

for all x_j, y_j in $\mathcal{M}$. This inequality is now lifted to ψ_1. Let $\mathcal{B}_0$ be the subalgebra of $\mathcal{B}$ spanned by the set of all projections in $\mathcal{B}$. Then the algebra $\mathrm{Alg}(\mathcal{M}, \mathcal{B}_0)$ is dense in $C^*(\mathcal{M}, \mathcal{B})$ so it is sufficient to prove that (1) holds with ψ_1 in place of ψ and x_j, y_j in $\mathrm{Alg}(\mathcal{M}, \mathcal{B}_0)$. As only finitely many elements are involved and each is generated by at most a finite number of projections in $\mathcal{B}_0$, there is a finite dimensional subalgebra $\mathcal{B}_1$ of $\mathcal{B}_0$ such that $x_1, \ldots, x_n, y_1, \ldots, y_n$ are in $\mathrm{Alg}(\mathcal{M}, \mathcal{B}_1)$. Let $q_1, \ldots, q_m$ be the minimal projections in $\mathcal{B}_1$. Then there are elements u_{jk} and v_{jk} in $\mathcal{M}$ such that

$$(2) \qquad x_j = \sum_{k=1}^{m} u_{jk} q_k \quad \text{and}$$

$$y_j = \sum_{k=1}^{m} v_{jk} q_k$$

for $1 \leq j \leq n$.

From the $\mathcal{B}$-module property of ψ_1 and the orthogonality of the projections q_j it follows that

$$\sum_j \psi_1(x_j, y_j) = \sum_k q_k \sum_j \psi(u_{jk}, v_{jk})$$

since the range of ψ is contained in $\mathcal{B}'$. The q_k's are orthogonal projections in the centre of $C^*(\mathcal{M}, \mathcal{B})$, so

$$(3) \qquad \left\| \sum_j \psi_1(x_j, y_j) \right\| \leq \max \left\{ \left\| \sum_j \psi(u_{jk}, v_{jk}) \right\| : 1 \leq k \leq m \right\}.$$

By inequality (1) for each k,

$$(4) \qquad \left\| \sum_j \psi(u_{jk}, v_{jk}) \right\| \leq 2\|\psi\| \left\| \sum u_{jk} u_{jk}^* \right\|^{\frac{1}{2}} \left\| \sum v_{jk}^* v_{jk} \right\|^{\frac{1}{2}}.$$

Also

$$\left\| \sum_j x_j x_j^* \right\| = \left\| \sum_{j,k,\ell} u_{jk} u_{j\ell}^* q_k q_\ell \right\|$$

$$= \left\| \sum_k \left(\sum_j u_{jk} u_{jk}^* \right) q_k \right\|$$

$$= \max \left\{ \left\| \sum_j u_{jk} u_{jk}^* \right\| : 1 \leq k \leq m \right\},$$

again using orthogonality of the q_k's. Combining this inequality and the corresponding one for $\left\| \sum_j y_j^* y_j \right\|$ with (3) and (4) yields

$$\left\| \sum_j \psi_1(x_j, y_j) \right\| \leq 2\|\psi\| \cdot \left\| \sum_j x_j x_j^* \right\|^{\frac{1}{2}} \left\| \sum_j y_j^* y_j \right\|^{\frac{1}{2}}$$

for all x_j, y_j in $\mathcal{M}$. This shows that ψ_1 lifts to a continuous linear map $\tilde{\psi}_1$ from $C^*(\mathcal{M}, \mathcal{B}) \otimes_h C^*(\mathcal{M}, \mathcal{B})$ into $B(H)$ with $\|\tilde{\psi}_1\| \leq 2\|\psi\|$. By construction ψ is a bilinear $C^*(\mathcal{A}, \mathcal{B})$-module map, since ψ is a bilinear $\mathcal{A}$-module map. By Theorem 5.3.12, the von Neumann algebra $(\mathcal{A} \cup \mathcal{B})''$ generated by $\mathcal{A} \cup \mathcal{B} = \mathcal{A} \cup J\mathcal{A}J$ is a masa in $B(H)$. Hence $(\mathcal{A} \cup \mathcal{B})''$ has a cyclic vector in H. Since $(\mathcal{A} \cup \mathcal{B})''$ is the ultraweak closure of $C^*(\mathcal{A}, \mathcal{B})$, this latter C^*-algebra also has a cyclic vector. Thus ψ_1 is completely bounded with $\|\psi_1\|_{cb} \leq 2\|\psi\|$ by Theorem 1.6.2. This shows that ψ is completely bounded with the same estimate.

5.4.7 Lemma. Let $\mathcal{M}$ be a type II_1 von Neumann algebra containing a hyperfinite von Neumann subalgebra $\mathcal{N}$ whose relative commutant $\mathcal{N}' \cap \mathcal{M}$ is the centre $\mathcal{Z}$ of $\mathcal{M}$. Let ϕ be a normal $\mathcal{N}$-module linear map from $\mathcal{M}$ into $B(H)$.
(1) Then ϕ is bounded on rows and columns of $\mathbf{M}_n(\mathcal{M})$ by $2\|\phi\|$; that is

$$\left\| \sum_1^n \phi(x_j)\phi(x_j)^* \right\| \leq 4\|\phi\|^2 \left\| \sum_1^n x_j x_j^* \right\|$$

and

$$\left\| \sum_1^n \phi(x_j)^* \phi(x_j) \right\| \leq 4\|\phi\|^2 \left\| \sum_1^n x_j^* x_j \right\|$$

for all $x_1, \ldots, x_n \in \mathcal{M}$ and $n \in \mathbb{N}$.

(2) If ϕ_n is the n-fold amplification of ϕ on $\mathbf{M}_n(\mathcal{M})$, then

$$\|\phi_n(x)\|^2 \leq 4\|\phi\|^2(\|x\|^2 + n\|\tau_n(x^* x)\|)$$

for all $n \in \mathbb{N}$ and $x \in \mathbf{M}_n(\mathcal{M})$, where τ_n is the normal centre-valued trace on $\mathbf{M}_n(\mathcal{M})$.

Proof. (1) Define $\psi \colon \mathcal{M} \times \mathcal{M} \to B(H)$ by

$$\psi(x, y) = \phi(x) \cdot \phi(y^*)^*$$

for all $x, y \in \mathcal{M}$. Then ψ is a normal bilinear map from $\mathcal{M} \times \mathcal{M}$ into $B(H)$ satisfying

$$\psi(xa, y) = \psi(x, ay)$$

for all $x, y \in \mathcal{M}$ and $a \in \mathcal{N}$, because ϕ is a right $\mathcal{N}$-module map. The first inequality in (1) now follows from Lemma 5.4.5(ii) by letting $y_j = x_j^*$ there. The second inequality may be obtained by applying the first to ϕ^* or using

$$\psi(x, y) = \phi(x^*)^* \phi(y)$$

in Lemma 5.4.5(ii).

(2) Let e_{ij} be the standard matrix units of $\mathbf{M}_n(\mathbf{C})$, regarded as elements of $\mathbf{M}_n(\mathcal{M})$ by identifying $\mathbf{M}_n(\mathbf{C})$ with $\mathbf{M}_n(\mathbf{C}) \otimes 1$. Let $\eta \in H^n$ with $\|\eta\| = 1$. Define the map ψ from $\mathbf{M}_n(\mathcal{M})$ into H^n by

$$\psi(x) = \phi_n(e_{11}x)\eta$$

for all $x \in \mathbf{M}_n(\mathcal{M})$. Now the linear map ψ is ultraweakly-weakly continuous, because ϕ is normal. Further $\|\psi\| \leq 2\|\phi\|$, because

$$\|\psi(x)\|^2 \leq \|\phi_n(e_{11}x)^* \phi_n(e_{11}x)\|$$
$$= \left\| \sum_j \phi(x_{1j})^* \phi(x_{1j}) \right\|$$
$$\leq 4\|\phi\|^2 \left\| \sum_j x_{1j}^* x_{1j} \right\|, \qquad \text{by (1),}$$
$$= 4\|\phi\|^2 \|(e_{11}x)^* e_{11}x\|$$
$$= 4\|\phi\|^2 \|e_{11}x\|^2$$
$$\leq 4\|\phi\|^2 \|x\|^2$$

for all $x \in \mathbf{M}_n(\mathcal{M})$.

By Corollary 5.4.3 there are normal states f and g on $\mathbf{M}_n(\mathcal{M})$ such that

$$\|\phi_n(e_{11}x)\eta\|^2 = \|\psi(x)\|^2 \leq 4\|\phi\|^2 (f(xx^*) + g(x^*x))$$

for all $x \in \mathbf{M}_n(\mathcal{M})$. The equation

$$x = \sum_i e_{i1} e_{11} e_{1i} x$$

and the $\mathbf{M}_n(\mathcal{N})$-modularity of ϕ_n lead to

$$\|\phi_n(x)\eta\|^2 = \left\| \sum e_{i1} \phi_n(e_{11}e_{1i}x) \right\|^2.$$

The orthogonality of $e_{i1}z$ and the estimate above now imply that

$$\|\phi_n(x)\eta\|^2 \leq \sum_i \|\phi_n(e_{11}e_{1i}x)\|^2$$

$$\leq 4\|\phi\|^2 \sum_i [f(e_{1i}xx^*e_{1i}) + g(x^*e_{i1}e_{1i}x)]$$

$$\leq 4\|\phi\|^2 [g(x^*x) + \sum_i f(e_{1i}xx^*e_{1i})].$$

The usual technique of introducing unitaries from the algebra with respect to which ϕ is a module map, now yields that for all $x \in M_n(\mathcal{M})$ and all unitaries $u \in M_n(\mathcal{N})$,

$$\|\phi_n(x)\eta\|^2 = \|u\phi_n(x)\eta\|^2$$

$$= \|\phi_n(ux)\eta\|^2$$

$$\leq 4g(x^*x) + 4\sum f(e_{1i}uxx^*u^*e_{i1}).$$

The hyperfinite algebra $M_n(\mathcal{N})$ is the weak closure of an increasing union of finite dimensional $*$-subalgebras $\mathcal{N}_\lambda$. Averaging over these as in (5.4.5) we obtain

$$\|\phi_n(x)\eta\|^2 \leq 4g(x^*x) + 4\sum f(e_{1i}\tau_n(xx^*)e_{i1})$$

$$\leq 4g(x^*x) + 4n\|\tau_n(x^*x)\|,$$

where τ_n is the centre-valued trace on $M_n(\mathcal{M})$. Since η was an arbitrary unit vector, this completes the proof.

Note that the state f on $M_n(\mathcal{M})$ may be assumed to be supported on $e_{11}M_n(\mathcal{M})e_{11}$ because it always occurs in the form $f(e_{11}yy^*e_{11})$; this shows that little information is lost in the last inequality.

The following lemma is used in Section 6.4 when we discuss the second cohomology group of a factor with property Γ.

5.4.8 Lemma. *Let $\mathcal{M}$ be a type II_1 von Neumann factor, let ξ be the unit tracial vector in $L^2(\mathcal{M})$, let $\mathcal{N}$ be a hyperfinite subfactor of $\mathcal{M}$ with trivial relative commutant $\mathcal{N}' \cap \mathcal{M} = \mathbb{C}1$, and let J be the involution on $L^2(\mathcal{M})$ given by $Jx = x^*$. If ϕ is a normal $\mathcal{N}$-module map from $\mathcal{M}$ into $(J\mathcal{N}J)'$, then*

$$\|\phi(x)\xi\| \leq 2\|\phi\|\,\|x\xi\|$$

for all $x \in \mathcal{M}$.

Proof. By Corollary 5.4.3 there are normal states f and g on $\mathcal{M}$ such that

$$\|\phi(x)\xi\|^2 \leq \|\phi\|^2(f(xx^*) + g(x^*x))$$

for all x in $\mathcal{M}$. If u is a unitary in $\mathcal{N}$, then

$$
\begin{aligned}
\|\phi(x)\xi\| &= \|uJuJ\phi(x)\xi\| \\
&= \|u\phi(x)JuJ\xi\| = \|u\phi(x)u^*\xi\| \\
&= \|\phi(uxu^*)\xi\|.
\end{aligned}
$$

Combining these two inequalities leads to

$$
\|\phi(x)\xi\| \le \|\phi\|(f(uxx^*u^*) + g(ux^*xu^*))
$$

for all x in $\mathcal{M}$ and all unitaries u in $\mathcal{N}$. Averaging, as in Lemma 5.4.5, over the unitary groups of an ascending chain of finite dimensional subalgebras of $\mathcal{N}$ leads to

$$
\begin{aligned}
\|\phi(x)\xi\| &\le \|\phi\|(f(tr(xx^*)1) + g(tr(x^*x)1)) \\
&= 2\|\phi\|tr(x^*x) \\
&= 2\|\phi\| \, \|x\xi\|
\end{aligned}
$$

as required.

5.5 Notes and Remarks

The results of Section 5.3 are due to Johnson, Kadison and Ringrose [JKR]. There are nice discussions of these methods and results in [Ri3] and [Ri6], which have influenced our account.

Popa's results are in [Pop1] and [Pop2]. There is a version of these results for general type II_1 von Neumann algebras with separable predual in [SiSm]. Note that examples due to Popa show the results fail without the hypothesis of separable predual (see [Pop1,Pop2]).

There is a proof of the non-commutative Grothendieck inequality for C^*-algebras in the book by Pisier [Pi2]; there are also proofs in [Pi1], [Ha3], [KS]. The normal bilinear form version (Corollary 5.4.2) is only available in [Ha3] so it is deduced from the usual non-commutative Grothendieck inequality. The two original accounts of the non-commutative Grothendieck inequality due to Pisier [Pi1] and Haagerup [Ha3] are recommended reading. The elimination of the anti-isomorphism terms from the Grothendieck inequality by averaging over a hyperfinite subalgebra with trivial relative commutant is in [ChS4] and [ChPSS]. The idea behind this technique is essentially contained in an earlier result by Christensen [Ch7] on similarities for type II_1 factors with property Γ.

6

Continuous Cohomology

6.1 Introduction

The continuous cohomology from a von Neumann algebra into itself is the principal topic of this chapter. The technical tools required in the proofs have been developed here, as in the last chapter, with a view that they may be useful in other von Neumann cohomology calculations.

The main technical result (6.2.1) of the second section is reminiscent of the theorems of Chapter 3, in that it says that continuous and completely bounded cohomology are equal under suitable hypotheses. This enables us to reduce continuous cohomology from a von Neumann algebra into dual normal modules to the completely bounded case when the algebra is type II_∞, III, or II_1 and stable under tensoring with $\mathcal{R}$, and the module is an operator space with multiplication. The possible difference between completely bounded and continuous cohomology thus lies in the algebras of type II_1 that are not stable under tensoring by $\mathcal{R}$.

By analogy with Lie groups, a Cartan subalgebra $\mathcal{A}$ of a von Neumann algebra $\mathcal{M}$ is a masa $\mathcal{A}$ whose unitary normalizer generates $\mathcal{M}$ as a von Neumann algebra. The automatic complete boundedness conditions of Section 1.6 and the averaged Haagerup–Pisier–Grothendieck inequality (5.4.5) show that, with the Cartan algebra hypothesis, suitable cocycles are completely bounded. The completely bounded cohomology results of Chapter 4 imply that

$$H^2(\mathcal{M}, \mathcal{M}) = H^3(\mathcal{M}, \mathcal{M}) = 0$$

if $\mathcal{M}$ has a Cartan subalgebra.

Property Γ for factors of type II_1 was introduced by Murray and von Neumann [MvN2] to show that there are two non-isomorphic factors of type II_1 on a separable Hilbert space. Using a result of Dixmier [Di1] on the existence of an abelian algebra with no minimal projections associated with an ultrapower of a type II_1 factor with property Γ, we show that $H^2(\mathcal{M}, \mathcal{M}) = 0$ for such a factor $\mathcal{M}$. Note that the Grothendieck inequality results (Section 5.4) are used here, which is an obstruction to extending this technique directly to the cases $n > 2$.

In Theorem 6.5.3 we show that if $H^n(\mathcal{M}, \mathcal{M}) = 0$ for all von Neumann algebras $\mathcal{M}$ on separable Hilbert spaces, then this holds for all von Neumann algebras $\mathcal{M}$.

6.2 Algebras Stable Under Tensor Products

In this section von Neumann algebras that are isomorphic to their von Neumann tensor product with $B(H)$ or with the type II_1 hyperfinite factor $\mathcal{R}$

are considered. Their ultraweak and ultraweak completely bounded cohomologies over a dual normal module are shown to be the same.

Recall that if ϕ is an n-linear map from an algebra $\mathcal{M}$ into an $\mathcal{M}$-module $\mathcal{V}$ and if ϕ is an $\mathcal{N}$-module mapping, then

$$\phi_k \colon \mathsf{M}_k(\mathcal{M})^n \to \mathsf{M}_k(\mathcal{V})$$

is an $\mathsf{M}_k(\mathcal{N})$ module mapping, directly from the "matrix multiplication" definition of ϕ_k (see Section 1.5).

6.2.1 Lemma. *Let $\mathcal{M} \subseteq B(H)$ be a von Neumann algebra having a subalgebra $\mathcal{N}$ isomorphic to the hyperfinite type II_1 factor $\mathcal{R}$ or to $B(H)$ such that $\mathcal{M} \cong \mathcal{M}\overline{\otimes}\mathcal{N}$. Let $\mathcal{V}$ be a subspace of $B(H)$ that is a dual normal $\mathcal{M}$-module. If ϕ is an n-linear normal $\mathcal{N}$-module map from $\mathcal{M}^n$ into $\mathcal{V}$, then ϕ is completely bounded with $\|\phi\|_{cb} = \|\phi\|$.*

Proof. In either of the cases, $\mathcal{N}$ isomorphic to $\mathcal{R}$ or to $B(H)$, there is an ultraweakly dense $*$-subalgebra $\mathcal{B}$ of $\mathcal{N}$ such that $\mathcal{B}$ is isomorphic to the infinite algebraic tensor product of M_2. Thus $\mathcal{B}$ is isomorphic to the inductive limit of the algebras

$$\mathsf{M}_2 \subseteq \mathsf{M}_2 \otimes \mathsf{M}_2 \subseteq \mathsf{M}_2 \otimes \mathsf{M}_2 \otimes \mathsf{M}_2 \subseteq \cdots .$$

Let $\mathcal{B}_j$ be the image of the jth algebra $\mathsf{M}_2^{\otimes j}$ in this inductive chain, and let $\mathcal{B}_j^c$ be the relative commutant of $\mathcal{B}_j$ in $\mathcal{B}$. Note that $\mathcal{B}_j^c$ is just the linear span of elements arising under the isomorphism from

$$1_1 \otimes \cdots \otimes 1_j \otimes x_{j+1} \otimes \cdots \otimes x_\ell,$$

where $x_i \in \mathsf{M}_2$. To simplify notation, elements arising from the isomorphism between $\mathcal{M}\overline{\otimes}\mathcal{N}$ and $\mathcal{M}$ are written as tensors. Let $\mathcal{M}_0$ be the von Neumann subalgebra of $\mathcal{M}$ isomorphic to $\mathcal{M} \otimes 1_{\mathcal{N}}$ under the isomorphism between $\mathcal{M}$ and $\mathcal{M}\overline{\otimes}\mathcal{N}$. Write $\mathcal{M}_0 \otimes \mathcal{B}$ for the $*$-subalgebra of $\mathcal{M}$ generated algebraically by $\mathcal{M}_0$ and $\mathcal{B}$. Note that $\mathcal{M}_0 \otimes \mathcal{B}$ is ultraweakly dense in $\mathcal{M}$.

To show that ϕ is completely bounded with $\|\phi\|_{cb} = \|\phi\|$ it is sufficient to show that

$$(1) \qquad\qquad \|\phi_{2^k}(x_1, \ldots, x_n)\| \le \|\phi\|$$

for all positive integers k and all $x_j \in \mathsf{M}_{2^k}(\mathcal{M})$ with $\|x_j\| \le 1$ for $1 \le j \le n$. By the Kaplansky density theorem applied to the $*$-subalgebra $\mathsf{M}_{2^k}(\mathcal{M}_0 \otimes \mathcal{B})$ of $\mathsf{M}_{2^k}(\mathcal{M})$, it is sufficient to prove (1) with $x_j \in \mathsf{M}_{2^k}(\mathcal{M}_0 \otimes \mathcal{B})$ and $\|x_j\| \le 1$ $(1 \le j \le n)$ because ϕ is normal. Identify

$$\mathsf{M}_{2^k}(\mathcal{M}) \quad \text{with} \quad \mathcal{M} \otimes \mathsf{M}_{2^k},$$

and

$$\mathcal{M}_0 \otimes \mathcal{B} \otimes \mathsf{M}_{2^k} \quad \text{with} \quad \mathcal{M}_0 \otimes (\mathcal{B}_t \otimes \mathsf{M}_{2^k} \otimes \mathcal{B}_s^c) \otimes \mathsf{M}_{2^k}$$

where $s = t + k$. There is a large value of t such that $x_j \in \mathsf{M}_{2^k}(\mathcal{M}_0 \otimes \mathcal{B}_t)$ so regard

$$(2) \qquad x_j \in \mathcal{M}_0 \otimes (\mathcal{B}_t \otimes \mathsf{C}1 \otimes \mathsf{C}1) \otimes \mathsf{M}_{2^k} \quad \text{for } 1 \le j \le n.$$

Automorphisms of matrix algebras are inner, so there is a unitary

$$(3) \qquad u \in \mathsf{C}1 \otimes (\mathsf{C}1 \otimes \mathsf{M}_{2^k} \otimes \mathsf{C}1) \otimes \mathsf{M}_{2^k} \subseteq \mathsf{C}1 \otimes \mathcal{B} \otimes \mathsf{M}_{2^k}$$

such that

$$(4) \qquad u(1 \otimes (1 \otimes \alpha \otimes 1) \otimes \beta)u^* = 1 \otimes (1 \otimes \beta \otimes 1) \otimes \alpha$$

for all $\alpha, \beta \in \mathsf{M}_{2^k}$. This is the inner automorphism on $\mathsf{M}_{2^k} \odot \mathsf{M}_{2^k}$ that interchanges the tensor factors. Let

$$y_j \otimes 1 = u x_j u^*$$

with $y_j \in \mathcal{M}_0 \otimes \mathcal{B}$ for $1 \le j \le n$ using (2) and (4). Now ϕ is a $(\mathsf{C}1 \otimes \mathcal{N})$-module map so ϕ_{2^k} is a $(\mathsf{C}1 \otimes \mathcal{N} \otimes \mathsf{M}_{2^k})$-module map (see the note before this lemma). These relations imply that

$$\begin{aligned}
\|\phi_{2^k}(x_1, \dots, x_n)\| &= \|u^* \phi_{2^k}(u x_1 u^*, \dots, u x_n u^*)u\| \\
&= \|\phi_{2^k}(y_1 \otimes 1, \dots, y_n \otimes 1)\| \\
&= \|\phi(y_1, \dots, y_n) \otimes 1\| \\
&\le \|\phi\|,
\end{aligned}$$

because

$$\phi_{2^k}(y_1 \otimes 1, \dots, y_n \otimes 1) = \phi(y_1, \dots, y_n) \otimes 1$$

and $\|y_j\| \le 1$ for $1 \le j \le n$. This proves Lemma 6.2.1.

6.2.2 Theorem. *Let $\mathcal{M}$ be a von Neumann algebra and let $\mathcal{V}$ be a subspace of $B(H)$ that is a dual normal $\mathcal{M}$-module. If the central direct summand $\mathcal{M}_{II_1}$ of type II_1 in $\mathcal{M}$ is isomorphic to its von Neumann algebra tensor product with the hyperfinite type II_1 factor $\mathcal{R}$, then*

$$H^n(\mathcal{M}, \mathcal{V}) \cong H_w^n(\mathcal{M}, \mathcal{V}) \cong H_{wcb}^n(\mathcal{M}, \mathcal{V}) \cong H_{cb}^n(\mathcal{M}, \mathcal{V}).$$

Proof. By Theorem 3.3.1 it is sufficient to prove

$$H_w^n(\mathcal{M}, \mathcal{V}) \cong H_{wcb}^n(\mathcal{M}, \mathcal{V}).$$

The von Neumann algebra decomposition

$$\mathcal{M} = \mathcal{M}_I \oplus \mathcal{M}_{II_1} \oplus \mathcal{M}_{II_\infty} \oplus \mathcal{M}_{III}$$

into its summands of types I, II_1, II_∞ and III induces a corresponding decomposition of cohomology groups (Corollary 3.3.8). The two groups $H^n_w(\mathcal{M}_I, \mathcal{V})$ and $H^n_{wcb}(\mathcal{M}_I, \mathcal{V})$ are both zero as type I algebras are hyperfinite. Now by the general structure theory of type II_∞ and type III von Neumann algebras, [Di2, KR4, T],

$$\mathcal{M}_{II_\infty} \cong \mathcal{M}_{II_\infty} \overline{\otimes} B(H) \text{ and } \mathcal{M}_{III} \cong \mathcal{M}_{III} \overline{\otimes} B(H).$$

Let $\mathcal{N}$ be the subalgebra of $\mathcal{M}_{II_\infty}$ which is isomorphic to $\mathbb{C}1 \otimes B(H)$. By Lemma 6.2.1,

$$H^n_w(\mathcal{M}_{II_\infty}, \mathcal{V}:/\mathcal{N}) \cong H^n_{wcb}(\mathcal{M}_{II_\infty}, \mathcal{V}:/\mathcal{N}).$$

It follows immediately from Theorem 3.4.1 that

$$H^n_w(\mathcal{M}_{II_\infty}, \mathcal{V}) \cong H^n_{wcb}(\mathcal{M}_{II_\infty}, \mathcal{V}).$$

An identical argument proves the corresponding isomorphism for $\mathcal{M}_{III}$ in place of $\mathcal{M}_{II_\infty}$, while the type II_1 case is handled by choosing $\mathcal{N}$ to be the isomorphic copy of $\mathbb{C}1 \otimes \mathcal{R}$ in $\mathcal{M}_{II_1} \cong \mathcal{M}_{II_1} \overline{\otimes} \mathcal{R}$. This completes the proof.

6.2.3 Theorem. *Let $\mathcal{M}$ be a von Neumann algebra in $B(H)$ and let $\mathcal{N}$ be an injective von Neumann subalgebra of $B(H)$ containing $\mathcal{M}$. If the central type II_1 direct summand $\mathcal{M}_{II_1}$ in $\mathcal{M}$ is isomorphic to its von Neumann algebra tensor product with the hyperfinite type II_1 factor, then*

$$H^n(\mathcal{M}, \mathcal{M}) = H^n(\mathcal{M}, \mathcal{N}) = 0, \quad (n \geq 1).$$

Proof. This follows directly from Theorem 6.2.2 and Theorems 4.2.6 and 4.3.1.

6.2.4 Problem. Let $\mathcal{M}$ be a type II_1 von Neumann algebra with separable predual and let $\mathcal{V}$ be a dual normal $\mathcal{M}$-module. If $\mathcal{M} \cong \mathcal{M} \otimes \mathbf{M}_k$ for some integer k greater than one, is

$$H^n_w(\mathcal{M}, \mathcal{V}) \cong H^n_{wcb}(\mathcal{M}, \mathcal{V})?$$

6.3 Cohomology with Cartan Subalgebras

Recall that a maximal abelian self-adjoint subalgebra $\mathcal{A}$ of a von Neumann algebra $\mathcal{M}$ is said to a *Cartan subalgebra* if its unitary normalizer

$$\mathcal{N}(\mathcal{A}) = \{u \in \mathcal{U}(\mathcal{M}): u\mathcal{A}u^* = \mathcal{A}\}$$

generates $\mathcal{M}$ as a von Neumann algebra. In this section we investigate the cohomology groups $H^n(\mathcal{M}, \mathcal{M})$, $n = 2, 3$, for von Neumann algebras with Cartan subalgebras. The main result is Theorem 6.3.1, where we show that both groups are zero.

At this time, it is unknown whether every type II_1 factor contains a Cartan subalgebra. It is possible that the von Neumann algebra generated by $\mathcal{N}(\mathcal{A})$ is $\mathcal{A}$ itself, in which case this masa is called *singular*. There are two natural masas in the group von Neumann algebra $VN(\mathsf{F}_2)$ of the free group F_2 on two generates g_1 and g_2. The first, generated by g_1 as a unitary in $VN(\mathsf{F}_2)$ (or its isomorphic copy generated by g_2), was shown to be singular by Popa [Pop1]. The second is the radial masa generated by the self-adjoint element

$$g_1 + g_2 + g_1^{-1} + g_2^{-1} \in VN(\mathsf{F}_2),$$

and this is also singular [R2].

The proof of the following theorem is based on all of our results up to this point. The underlying general idea is as follows. If $\mathcal{N}$ is hyperfinite, then averaging and normal cohomology results enable us to pass from $H^n(\mathcal{M}, \mathcal{M})$ to $H_w^n(\mathcal{M}, \mathcal{M}:/\mathcal{N})$ – a step that works for any dual normal $\mathcal{M}$-module. The corresponding completely bounded cohomology space $H_{wcb}^n(\mathcal{M}, \mathcal{M}:/\mathcal{N})$ is zero so it is sufficient to show that, with the Cartan subalgebra hypothesis,

$$H_w^n(\mathcal{M}, \mathcal{M}:/\mathcal{N}) \cong H_{wcb}^n(\mathcal{M}, \mathcal{M}:/\mathcal{N}).$$

This is where Sections 1.6 and 5.4 enter the picture, via Lemma 5.4.6.

6.3.1 Theorem. *Let $\mathcal{M}$ be a type II_1 von Neumann algebra with separable predual. If $\mathcal{M}$ has a Cartan subalgebra, then*

$$H^2(\mathcal{M}, \mathcal{M}) = H^3(\mathcal{M}, \mathcal{M}) = 0.$$

Proof. The proof is only given for a factor because of (5.3.10). Let $\mathcal{A}$ be a Cartan subalgebra in $\mathcal{M}$, and let $\mathcal{N}$ be the hyperfinite von Neumann subalgebra of $\mathcal{M}$ containing $\mathcal{A}$ with $\mathcal{N}' \cap \mathcal{M} = \mathcal{Z}$, given by Theorem 5.3.10. Let tr be a faithful normal trace on $\mathcal{M}$, let $H = L^2(\mathcal{M}, tr)$ and let J be the standard involution on H. We note that $x \to JxJ$ is a conjugate linear anti-isomorphism of $\mathcal{M}$ onto $\mathcal{M}'$, and so $J\mathcal{A}J$ is a masa in $\mathcal{M}'$.

By Theorems 3.4.1 and 4.3.1

$$H^n(\mathcal{M}, \mathcal{M}) = H_w^n(\mathcal{M}, \mathcal{M}:/\mathcal{N})$$

for all $n \geq 1$, so it is sufficient to show that

$$H_w^n(\mathcal{M}, \mathcal{M}:/\mathcal{N}) = 0$$

for $n = 2$ and 3. By Theorem 4.3.1 $H_{wcb}^n(\mathcal{M}, \mathcal{M}) = 0$ so it is sufficient to show that if $\phi \in \mathcal{L}_w^n(\mathcal{M}, \mathcal{M}:/\mathcal{N})$ with $\partial\phi = 0$, then ϕ is completely bounded. By Corollary 5.2.4, there exists

$$\psi \in \mathcal{L}_w^{n-1}(C^*(\mathcal{M}, JAJ), (JAJ)':/C^*(\mathcal{N}, JAJ))$$

such that $\phi = \partial\psi$ when restricted to $\mathcal{M}$. If $n = 2$, then ψ satisfies the hypotheses of Lemma 5.4.6 and hence is completely bounded. If $n = 3$, then ψ satisfies the hypotheses of Lemma 5.4.6 so ψ, and hence $\phi = \partial\psi$, is completely bounded. This proves the theorem.

6.3.2 Remark. The proof of Lemma 5.4.6, and so Theorem 6.3.1, depends on the non-commutative Grothendieck inequality. Currently this is the only known route to proving that $H^3(\mathcal{M}, \mathcal{M}) = 0$ if $\mathcal{M}$ is a type II_1 von Neumann algebra with Cartan subalgebra. The corresponding result $H^2(\mathcal{M}, \mathcal{M}) = 0$ need not depend on the non-commutative Grothendieck inequality. When $n = 2$, ψ is a linear map and the one-variable version of Theorem 1.6.1 may be used to deduce that ψ is completely bounded.

Note that the existence of a cyclic vector for the von Neumann algebra $(\mathcal{A} \cup JAJ)''$ is crucial in the proof. It would be sufficient to have hypotheses that implied that the algebra $(\mathcal{N} \cup JAJ)''$ satisfied condition (ii) of Theorem 1.6.2 because then $C^*(\mathcal{N}, JAJ)$ would also satisfy this condition and ψ would be completely bounded.

6.4 Cohomology for Factors with Property Γ

Murray and von Neumann [MvN2] introduced property Γ to show that there were two non-isomorphic type II_1 factors on a separable Hilbert space. Throughout this section $\mathcal{M}$ denotes a type II_1 factor with normalized trace tr and $\|x\|_2 = tr(x^*x)$ for all $x \in \mathcal{M}$.

6.4.1 Definition. The factor $\mathcal{M}$ is said to have property Γ if for each $\varepsilon > 0$, $n \in \mathsf{N}$ and $x_1, \ldots, x_n \in \mathcal{M}$ there is a unitary u in $\mathcal{M}$ with $\|ux_j - x_j u\|_2 < \varepsilon$, and $tr(u) = 0$.

The property will be used here via an ultrapower of the algebra. Let ω be a free ultrafilter on N, so that ω corresponds to a character in $\beta\mathsf{N}\backslash\mathsf{N}$, in the

Stone–Čech compactification of N but not in N, and let $\lim_{\omega} \alpha_n$ be $\omega((\alpha_n))$ for (α_n) in $\ell^\infty(N)$. Let $\mathcal{I}_\omega$ be the closed two-sided ideal in $\ell^\infty(N, \mathcal{M}) = \ell^\infty(N)\overline{\otimes}\mathcal{M}$ of those sequences (x_n) with $\lim_{\omega} \|x_n\|_2 = 0$. This is a maximal ideal in $\ell^\infty(N, \mathcal{M})$, and the quotient $\mathcal{M}^\omega \equiv (\ell^\infty(N)\overline{\otimes}\mathcal{M})/\mathcal{I}_\omega$ is a type II_1 factor [Co2, McD]. The normalized trace on the quotient is defined by

$$\tau((x_n) + \mathcal{I}_\omega) = \lim_{\omega} tr(x_n).$$

Let H^ω denote the Hilbert space $L^2(\mathcal{M}^\omega, \tau)$ and let $\mathcal{M}^\omega$ act standardly on $L^2(\mathcal{M}^\omega, \tau)$ by left multiplication. Its commutant is $\mathcal{M}^\omega$ acting by right multiplication. Embed $\mathcal{M}$ into $\mathcal{M}^\omega$ by mapping an element x in $\mathcal{M}$ to the constant sequence $(x, x, \ldots) + \mathcal{I}_\omega$; denote the image of $\mathcal{M}$ by $\overline{\mathcal{M}}$, which is a von Neumann subalgebra of $\mathcal{M}^\omega$ isomorphic to $\mathcal{M}$. Let $\mathcal{M}_\omega$ denote the relative commutant $\overline{\mathcal{M}}' \cap \mathcal{M}^\omega$ of $\overline{\mathcal{M}}$ in $\mathcal{M}^\omega$. Now Dixmier's characterization [Di1, Proposition 1.10] is that there is an abelian von Neumann algebra $\mathcal{A}$ contained in $\mathcal{M}_\omega$ with no minimal projections. This abelian algebra $\mathcal{A}$ provides us with the ability to subdivide the identity by projections with arbitrarily small trace.

6.4.2 Theorem. *If $\mathcal{M}$ is a type II_1 von Neumann factor with property Γ and separable predual, then $H^2(\mathcal{M}, \mathcal{M}) = 0$.*

Proof. The idea is to show that a suitably modified cocycle is completely bounded, and hence is a coboundary by Theorem 4.3.1. This underlying idea is similar to the Cartan case. However, having transferred the problem to the ultrapower algebra $\mathcal{M}^\omega$, an additional averaging over an abelian subalgebra $\mathcal{A}$ of $\mathcal{M}^\omega \cap \overline{\mathcal{M}}'$ is required to make the cocycle $\mathcal{A}$-modular.

Let ω be a free ultrafilter on N and let $\mathcal{N}$ be a hyperfinite von Neumann subalgebra of $\mathcal{M}$ with $\mathcal{N}' \cap \mathcal{M} = C1$ (Theorem 5.3.7). By Theorem 3.4.1 it is sufficient to show that $H^2_w(\mathcal{M}, \mathcal{M}: /\mathcal{N})$ is zero. Let ϕ be a normal bilinear $\mathcal{N}$-module map from $\mathcal{M} \times \mathcal{M}$ into $\mathcal{M}$ with $\partial\phi = 0$. Now the algebra $J\mathcal{N}J$ is hyperfinite and contained in $J\mathcal{M}J = \mathcal{M}'$, so $(J\mathcal{N}J)'$ is a hyperfinite algebra containing $\mathcal{M}$. By Corollary 5.2.4, there is a normal $\mathcal{N}$-module map ψ from $\mathcal{M}$ into $(J\mathcal{N}J)'$ with $\phi = \partial\psi$.

To lift ϕ to the ultrapower $\mathcal{M}^\omega$ of $\mathcal{M}$ it is necessary to show that $\phi(\mathcal{I}_\omega, \ell^\infty(N, \mathcal{M})) \subseteq \mathcal{I}_\omega$ and $\phi(\ell^\infty(N, \mathcal{M}), \mathcal{I}_\omega) \subseteq \mathcal{I}_\omega$. The hypotheses on ϕ also apply to the cocycle ϕ^*, where

$$\phi^*(x, y) = \phi(y^*, x^*)^*$$

for all $x, y \in \mathcal{M}$. Note that

$$\phi^*(\mathcal{I}_\omega, \ell^\infty(N, \mathcal{M})) = \phi(\ell^\infty(N, \mathcal{M}), \mathcal{I}_\omega)^*$$

so it is sufficient to show that if $(x_n) \in \ell^\infty(\mathsf{N}, \mathcal{M})$ and $(y_n) \in \mathcal{I}_\omega$, then $(\phi(x_n, y_n)) \in \mathcal{I}_\omega$. Now

$$
\begin{aligned}
\|\phi(x_n, y_n)\xi\| &= \|x_n\psi(y_n)\xi - \psi(x_n y_n)\xi + \psi(x_n)y_n\xi\| \\
&\leq 2\|x_n\|\,\|\psi\|\,\|y_n\xi\| + 2\|\psi\| \cdot \|x_n y_n \xi\| \\
&\quad + \|\psi\|\,\|x_n\|\,\|y_n\xi\| \\
&= 5C\|\psi\|\,\|y_n\xi\|
\end{aligned}
$$

by Lemma 5.4.7 where $\|x_n\| \leq C$ for all n. Taking the limit over the ultrafilter ω proves that $(\phi(x_n, y_n))$ is in $\mathcal{I}_\omega$. Hence ϕ lifts to a bilinear map ϕ^ω from $\mathcal{M}^\omega \times \mathcal{M}^\omega$ into $\mathcal{M}^\omega$ with $\partial\phi^\omega = 0$, by defining

$$
\phi^\omega((x_n) + \mathcal{I}_\omega, (y_n) + \mathcal{I}_\omega) = (\phi(x_n, y_n)) + \mathcal{I}_\omega
$$

for all $(x_n), (y_n)$ in $\ell^\infty(\mathsf{N}, \mathcal{M})$.

By Lemma 5.4.7,

$$
(1) \qquad \|\psi_n(x)\|^2 \leq 4\|\psi\|^2(\|x\|^2 + ntr(x^*x))
$$

for all x in $\mathsf{M}_n(\mathcal{M})$, where ψ_n is the n-fold amplification of ψ. Since amplification commutes with the coboundary operator ∂, it is clear that $\partial\psi_n = \phi_n$. Thus the equation

$$
\phi_n(x, y) = x\psi_n(y) - \psi_n(xy) + \psi_n(x)y
$$

implies that

$$
(2) \qquad \|\phi_n(x, y)\|^2 \leq 36\|\psi\|^2(\|x\|^2 + ntr(x^*x))(\|y\|^2 + ntr(y^*y))
$$

for all $x, y \in \mathsf{M}_n(\mathcal{M})$ by (1). Here we have used the estimate

$$
\begin{aligned}
\|xy\|^2 + ntr((xy)^*xy) &\leq \|x\|^2\|y\|^2 + ntr(y^*y)\|x\|^2 \\
&\leq (\|x\|^2 + ntr(x^*x))(\|y\|^2 + ntr(y^*y)).
\end{aligned}
$$

This inequality now lifts via $\ell^\infty(\mathsf{N}, \mathcal{M})$ to $\mathcal{M}^\omega$ and ϕ^ω. Hence

$$
(3) \qquad \|(\phi^\omega)_n(x, y)\|^2 \leq 36\|\psi\|^2(\|x\|^2 + ntr(x^*x))^{\frac{1}{2}}(\|y\|^2 + ntr(y^*y))^{\frac{1}{2}}
$$

for all x, y in $\mathsf{M}_n(\mathcal{M}^\omega)$.

Now Dixmier's result [Di1, Proposition 1.10] on factors with property Γ implies that there is an abelian von Neumann subalgebra $\mathcal{A}$ of $\mathcal{M}_\omega = \overline{\mathcal{M}' \cap \mathcal{M}^\omega}$ with no minimal projections. The averaging methods of Chapter 3 yield a map $\gamma \colon \mathcal{M}^\omega \to \mathcal{M}^\omega$ such that $\chi = \phi^\omega - \partial\gamma$ is a normal bilinear $\mathcal{A}$-module map from $\mathcal{M}^\omega \times \mathcal{M}^\omega$ into $\mathcal{M}^\omega$. The crucial point is that inequality

(3) holds for χ in place of ϕ^ω with the constant 36 replaced by a larger constant ($= 900$). The idea of the proof of this is straightforward as at each stage of the construction of γ, in Sections 3.2 and 3.3, averaging is used. This process preserves the inequality, though the constant is increased by an integer multiple each time as several terms occur – see Lemma 3.4.2.

The full details of this rather tedious procedure are not given. Note that the process has to be gone through twice. Once to lift the adjusted ϕ^ω to be normal, which is done by averaging over a two dimensional subalgebra of $(\mathcal{M}^\omega)^{**}$ – see the proof of (3.3.4). The second is the averaging of the resulting cocycle over the unitary group of $\mathcal{A}$ to ensure the result is an $\mathcal{A}$-module map – each introduces an additional factor of 5 (see 3.4.2). Hence suppose that

$$(4) \qquad \|\chi_n(x,y)\|^2 \leq K(\|x\|^2 + ntr(x^*x))^{\frac{1}{2}} \cdot (\|y\|^2 + ntr(y^*y))^{\frac{1}{2}}$$

for all $x, y \in \mathbf{M}_n(\mathcal{M})$, where $K \leq \|\psi\| \cdot 900$.

Since $\mathcal{A}$ is contained in $\overline{\mathcal{M}}' \cap \mathcal{M}^\omega$, for each $a \in \mathcal{A} \otimes 1 \subseteq \mathcal{M}^\omega \otimes \mathbf{M}_n$, $ax = xa$, $ay = ya$ and

$$\chi_n(x,y)a = \chi_n((x,ya) = \chi_n(x,ay) = \chi_n(xa,y) = \chi_n(ax,y) = a\chi_n(x,y)$$

for all $x, y \in \mathbf{M}_n(\mathcal{M})$. This shows that χ_n maps $\mathbf{M}_n(\overline{\mathcal{M}}) \times \mathbf{M}_n(\overline{\mathcal{M}})$ into the commutant of $I \otimes \mathcal{A}$ in $\mathbf{M}_n(\mathcal{M}^\omega)$. Since $\mathcal{A}$ has no minimal projections the identity in $I \otimes \mathcal{A}$ may be decomposed as a finite sum of projections $1 = \sum_1^k p_j$ with $tr(p_j) \leq 1/n$ for each j. Now for each $z \in \mathbf{M}_n(\mathcal{M}^\omega) \cap (I \otimes \mathcal{A})'$,

$$\|z\| = \left\| \sum_1^k p_j z p_j \right\|$$
$$= \max\{\|p_j z p_j\| \colon 1 \leq j \leq k\}.$$

Let x and y be in $\mathbf{M}_n(\overline{\mathcal{M}})$ with $\|x\| \leq 1$ and $\|y\| \leq 1$, and let p be that p_j with

$$\|\chi_n(x,y)\| = \|p_j\chi_n(x,y)p_j\| = \|p\chi_n(x,y)p\|.$$

Now $p\chi_n(x,y)p = \chi_n(px,yp)$ so that

$$\|\chi_n(x,y)\| = \|p\chi_n(x,y)p\| = \|\chi_n(px,py)\|$$
$$\leq K(\|px\|^2 + ntr(pxx^*p))^{\frac{1}{2}} \cdot (\|py\|^2 + ntr(pyy^*p))^{\frac{1}{2}}$$
$$\text{(by (4) since } py = yp)$$
$$\leq K(\|x\|^2 + \|xx^*\|ntr(p))^{\frac{1}{2}} \cdot (\|y\|^2 + \|yy^*\|ntr(p))^{\frac{1}{2}}$$
$$= 2K\|x\| \cdot \|y\|$$

since $tr(p) \leq 1/n$.

This implies that χ restricted to $\overline{\mathcal{M}}$ is completely bounded with completely bounded norm at most $2K$. Since $\mathcal{M}^\omega$ is a type II_1 factor, there is a conditional expectation E from $\mathcal{M}^\omega$ onto $\overline{\mathcal{M}}$. Further E is an $\overline{\mathcal{M}}$-module map and is completely bounded because it is completely positive. Hence on $\overline{\mathcal{M}} \times \overline{\mathcal{M}}$,

$$E\chi = E\phi - E\partial\gamma = \phi - \partial E\gamma$$

and on $\overline{\mathcal{M}} \times \overline{\mathcal{M}} \times \overline{\mathcal{M}}$,

$$\partial E\chi = E\partial\chi = 0.$$

Now $\overline{\mathcal{M}}$ is isomorphic to $\mathcal{M}$ so that $E\chi = \partial\theta$ for some completely bounded map θ on $\overline{\mathcal{M}}$. Thus $\phi = \partial(\theta + E\gamma)$ so $H^2(\mathcal{M}, \mathcal{M}) = 0$ as required.

6.5 Separable Preduals

In this section we show that it suffices to study cohomology in the setting of a von Neumann algebra on a separable Hilbert space; either $H^n(\mathcal{M}, \mathcal{M}) = 0$ for all von Neumann algebras, or this fails for a separably acting one.

6.5.1 Lemma. *Let $n \geq 2$ be fixed, and let $\mathcal{C}$ be a class of von Neumann algebras which is closed under countable direct sums, and whose elements satisfy $H^n(\mathcal{M}, \mathcal{M}) = 0$. Then there is a constant Λ_n such that every n-cocycle ϕ on an element of the class is an n-coboundary $\partial\psi$ with $\|\psi\| \leq \Lambda_n\|\phi\|$.*

Proof. Suppose that no such constant exists. Then, for each integer $k \geq 1$ we may find von Neumann algebras $\mathcal{M}_k \in \mathcal{C}$, and n-cocycles ϕ_k on $\mathcal{M}_k$ of unit norm, such that $\|\psi_k\| \geq k$ whenever $\phi_k = \partial\psi_k$. By hypothesis $\mathcal{M} = \bigoplus_{k=1}^{\infty} \mathcal{M}_k \in \mathcal{C}$, and contains a copy of ℓ_∞ by taking direct sums of the units in the $\mathcal{M}_k$'s. By Theorems 3.3.1 and 3.4.1, we may assume multimodularity with respect to this abelian subalgebra for all cocycles and coboundaries on $\mathcal{M}$. Then it is clear that $\phi = \bigoplus_{k=1}^{\infty} \phi_k$ is a non-trivial cocycle on $\mathcal{M}$ and so $H^n(\mathcal{M}, \mathcal{M}) \neq 0$, a contradiction.

We have in mind two such classes of von Neumann algebras:

(i) those which have faithful representations on separable Hilbert spaces,
(ii) those which have unital separable ultraweakly dense C^*-subalgebras.

To see that the second class is closed under countable direct sums, observe that the c_0-sum of separable ultraweakly dense C^*-subalgebras (with unit adjoined) is separable and ultraweakly dense in the direct sum.

6.5.2 Lemma. *Let* $\mathcal{M} \subseteq B(H)$ *be a von Neumann algebra which contains a separable ultraweakly dense C^*-subalgebra $\mathcal{D}$. Then there exist disjoint central projections $\{p_\alpha\}$, $\sum_\alpha p_\alpha = 1$, such that each $\mathcal{M}p_\alpha$ can be faithfully represented on a separable Hilbert space.*

Proof. If ξ is a non-zero vector in H then $H_1 = \overline{\mathcal{D}\xi}$, the norm closed cyclic subspace generated by $\mathcal{D}$ and ξ, is separable and the associated projection q lies in $\mathcal{D}' = \mathcal{M}'$. Thus there is a normal representation $m \to mq$ of $\mathcal{M}$ on H_1. The kernel is an ultraweakly closed ideal in $\mathcal{M}$, so has the form $\mathcal{M}(1 - p)$ for a central projection p. Then this representation is faithful on $\mathcal{M}p$.

Now let $\{p_\alpha\}$ be a maximal family of disjoint central projections such that each $\mathcal{M}p_\alpha$ has a faithful normal representation on a separable Hilbert space. The above construction shows that this collection is non-empty. If $p_0 = \sum_\alpha p_\alpha < 1$, then consider $\mathcal{M}(1 - p_0)$ with ultraweakly dense separable C^*-subalgebra $\mathcal{D}(1 - p_0)$. As before there is a non-zero central projection $p \in \mathcal{Z}(1 - p_0)$ such that $\mathcal{M}(1 - p_0)p = \mathcal{M}p$ can be faithfully represented on a separable Hilbert space. This contradicts the maximality of $\{p_\alpha\}$, and so we must have $\sum p_\alpha = 1$ as required.

6.5.3 Theorem. *Let $n \geq 2$ be a fixed integer and suppose that $H^n(\mathcal{M}, \mathcal{M}) = 0$ for all von Neumann algebras on separable Hilbert spaces. Then $H^n(\mathcal{M}, \mathcal{M}) = 0$ for all von Neumann algebras.*

Proof. We will only discuss the case $n = 2$; the other cases may be proved in a similar manner but the details are more complicated.

Let $\mathcal{M} \subseteq B(H)$ be a von Neumann algebra, and consider a cocycle

$$\phi \colon \mathcal{M} \times \mathcal{M} \to \mathcal{M},$$

which we may assume to be separately normal in each variable. To each finite subset F of $\mathcal{M}$ we will associate a unital separable C^*-subalgebra $\mathcal{D}_F$ of $\mathcal{M}$ with the following properties:

(i) $F \subseteq \mathcal{D}_F$,
(ii) if $F \subseteq G$ then $\mathcal{D}_F \subseteq \mathcal{D}_G$,
(iii) ϕ maps $\mathcal{D}_F \times \mathcal{D}_F$ into $\mathcal{D}_F$.

Given a finite subset F of $\mathcal{M}$, let $\mathcal{D}_0$ be the unital separable C^*-algebra generated by F and then define inductively $\mathcal{D}_n$ to be the separable C^*-algebra generated by $\mathcal{D}_{n-1}$ and the image of ϕ restricted to $\mathcal{D}_{n-1} \times \mathcal{D}_{n-1}$. Then let $\mathcal{D}_F$ be the norm closure of $\bigcup_{n=0}^{\infty} \mathcal{D}_n$. By construction $\mathcal{D}_F$ is separable, ϕ maps $\mathcal{D}_n \times \mathcal{D}_n$ into $\mathcal{D}_{n+1}$ and so maps $\mathcal{D}_F \times \mathcal{D}_F$ into $\mathcal{D}_F$, and $\mathcal{D}_F$ is the smallest unital C^*-algebra $\mathcal{A}$ with the property that $F \subseteq \mathcal{A}$ and ϕ maps $\mathcal{A} \times \mathcal{A}$ to $\mathcal{A}$. It then follows that $\mathcal{D}_F \subseteq \mathcal{D}_G$ whenever $F \subseteq G$. Now let $\mathcal{M}_F$ be the ultraweak closure of $\mathcal{D}_F$ in $\mathcal{M}$. By normality of ϕ, its restriction ϕ_F to $\mathcal{M}_F \times \mathcal{M}_F$ maps into $\mathcal{M}_F$. Lemma 6.5.2 allows us to decompose $\mathcal{M}_F$ as $\bigoplus_\alpha \mathcal{M}_{Fp_\alpha}$ where each $\mathcal{M}_{Fp_\alpha}$ acts on a separable Hilbert space and the p_α's are central projections in $\mathcal{M}_F$. By (3.1.1) there exists an absolute constant C and a normal map $\theta_F \colon \mathcal{M}_F \to \mathcal{M}_F$ such that $\phi_F - \partial\theta_F$ is multimodular with respect to the centre of $\mathcal{M}_F$ and

$$\|\phi_F - \partial\theta_F\| \leq C\|\phi_F\| \leq C\|\phi\|,$$
$$\|\theta_F\| \leq C\|\phi_F\| \leq C\|\phi\|.$$

Then $\phi_F - \partial\theta_F$ is a cocycle on each $\mathcal{M}_{Fp_\alpha}$, so by the hypothesis and Lemma 6.5.1, there exists a constant Λ_2 and there exist normal maps $\mu_\alpha \colon \mathcal{M}_{Fp_\alpha} \to \mathcal{M}_{Fp_\alpha}$,

$$\|\mu_\alpha\| \leq \Lambda_2\|\phi_F - \partial\theta_F\| \leq \Lambda_2(C+1)\|\phi\|,$$

such that

$$\phi_F - \partial\theta_F = \partial\left(\bigoplus_\alpha \mu_\alpha\right).$$

Let $\psi_F \colon \mathcal{M}_F \to \mathcal{M}_F$ be $\theta_F + \bigoplus_\alpha \mu_\alpha$. Then

$$\|\psi_F\| \leq \|\theta_F\| + \sup_\alpha \|\mu_\alpha\|$$
$$\leq C\|\phi\| + \Lambda_2(C+1)\|\phi\|$$
$$= C'\|\phi\|$$

where $C' = C + \Lambda_2(C+1)$.

Let $\mathcal{S}$ be the set of finite subsets F of $\mathcal{M}$, ordered by inclusion, and choose a state ω on the algebra $B(\mathcal{S})$ of bounded functions on $\mathcal{S}$, such that

$$\omega(f) = \lim_F f(F)$$

whenever this limit exists. For $x \in \mathcal{M}$, $\xi, \eta \in H$, define $f_{x,\xi,\eta} \in B(\mathcal{S})$ by

$$f_{x,\xi,\eta}(F) = \begin{cases} \langle \psi_F(x)\xi, \eta \rangle & \text{if } x \in F, \\ 0 & \text{otherwise.} \end{cases}$$

For $x \in \mathcal{M}$,

$$[\xi, \eta] = \omega(f_{x,\xi,\eta})$$

defines a bounded sesquilinear form on $H \times H$. Thus there is a bounded linear operator $\psi(x)$ such that

$$\omega(f_{x,\xi,\eta}) = \langle \psi(x)\xi, \eta \rangle,$$

and it is easy to obtain the estimate $\|\psi(x)\| \leq C'\|x\|$. If F contains x, y, and $x + y$ then

$$f_{x,\xi,\eta}(F) + f_{y,\xi,\eta}(F) = f_{x+y,\xi,\eta}(F),$$

and so

$$\lim_F (f_{x,\xi,\eta}(F) + f_{y,\xi,\eta}(F) - f_{x+y,\xi,\eta}(F)) = 0.$$

Thus

$$\omega(f_{x,\xi,\eta}) + \omega(f_{y,\xi,\eta}) = \omega(f_{x+y,\xi,\eta}),$$

showing that

$$\psi(x + y) = \psi(x) + \psi(y).$$

Similarly ψ respects multiplication by scalars. For $t \in \mathcal{M}'$, it is clear from the definition that

$$f_{x,t\xi,\eta} = f_{x,\xi,t^*\eta},$$

and so

$$\begin{aligned}
\langle \psi(x)t\xi, \eta \rangle &= \langle \psi(x)\xi, t^*\eta \rangle \\
&= \langle t\psi(x)\xi, \eta \rangle.
\end{aligned}$$

This shows that $\psi(x) \in (\mathcal{M}')' = \mathcal{M}$. Now consider a fixed but arbitrary pair of elements $x, y \in \mathcal{M}$. If F contains $\{x, y, xy\}$, then

$$\begin{aligned}
\langle \phi(x, y)\xi, \eta \rangle &= \langle \phi_F(x, y)\xi, \eta \rangle \\
&= \langle x\psi_F(y)\xi, \eta \rangle - \langle \psi_F(xy)\xi, \eta \rangle + \langle \psi_F(x)y\xi, \eta \rangle \\
&= f_{y,\xi,x^*\eta}(F) - f_{xy,\xi,\eta}(F) + f_{x,y\xi,\eta}(F)
\end{aligned}$$

and so

$$\begin{aligned}
\langle \phi(x, y)\xi, \eta \rangle &= \omega(f_{y,\xi,x^*\eta}) - \omega(f_{xy,\xi,\eta}) + \omega(f_{x,y\xi,\eta}) \\
&= \langle x\psi(y)\xi, \eta \rangle - \langle \psi(xy)\xi, \eta \rangle + \langle \psi(x)y\xi, \eta \rangle,
\end{aligned}$$

for $\xi, \eta \in H$. It follows that $\phi = \partial \psi$ and $H^2(\mathcal{M}, \mathcal{M}) = 0$.

6.6 Notes and Remarks

The results of Section 6.2 are taken from [ChES]. It would be interesting to know if the continuous (Hochschild) cohomology of a type II_1 von Neumann algebra $\mathcal{M}$ with separable predual is equal to the completely bounded cohomology if $\mathcal{M} \cong \mathcal{M} \otimes \mathbf{M}_n$ for some $n \geq 2$. What happens if $\mathcal{M}$ is a type II_1 factor with full or trivial fundamental group?

Johnson [J5] constructed a type II_1 von Neumann algebra that was not injective and for which $H^2(\mathcal{M}, \mathcal{M}) = 0$. An examination of the construction shows that the algebra has a Cartan subalgebra. This example of Johnson's is a precursor to the results of Section 6.3 and is implied by them. The results of Section 6.3 are from [ChPSS] where the transition from factors to the general case is handled by direct integral theory; some of the ideas and methods evolved from [PoS].

Section 6.4 is taken from [ChS4], though the ideas in this section go back to Christensen's research on derivations and similarities [Ch3], [Ch7].

Section 6.5 is probably well known to experts, but has not appeared in the literature.

7

Stability of Products

7.1 Introduction

This chapter presents some applications of cohomology to the structure theory of von Neumann algebras. Much of this material applies to arbitrary Banach algebras without modification, and so we have stated many of the results in full generality.

Section 7.2 explores the relationship between the first cohomology group and the principal component in the automorphism group of a Banach algebra. The main result (7.2.5) shows that, under the hypothesis $H^1(\mathcal{A}, \mathcal{A}) = 0$, any automorphism in the principal component of $\mathrm{Aut}(\mathcal{A})$ is not only inner but is implemented by an element of the principal component of the group $\mathcal{A}^{-1}$ of invertible elements in $\mathcal{A}$.

Section 7.3 contains an implicit function theorem for Fréchet differentiable maps on Banach spaces, which we apply to questions of stability in Section 7.4. Here we present two main results. If $\lambda\colon \mathcal{A} \to \mathcal{A}$ is an invertible linear map then $(x, y) \to \lambda^{-1}(\lambda(x)\lambda(y))$ defines a new associative product on $\mathcal{A}$. Under the hypothesis of vanishing second and third cohomology groups, we show that any associative product sufficiently close to the original one has such a form, and λ can be chosen to be close to the identity map. The second stability result is a consequence of this: any von Neumann algebra sufficiently close to an injective von Neumann algebra must itself be injective, and the two algebras are isomorphic.

7.2 Principal Component of the Automorphism Group

A continuous derivation $\delta\colon \mathcal{A} \to \mathcal{A}$ on any Banach algebra $\mathcal{A}$ may be exponentiated, and a standard calculation shows that e^δ is an automorphism. When δ is inner and implemented by $a \in \mathcal{A}$, the associated automorphism is inner and is implemented by e^a. It is natural to ask whether all automorphisms can be obtained as exponentials of derivations. This fails in general, but is correct under additional hypotheses. The next result is outside the scope of these notes and so we refer to [ZeM] for the proof (see also the appendix to the English translation of [Di2]).

7.2.1 Theorem. *Let $\mathcal{A}$ be a Banach algebra and let α be a bounded automorphism of $\mathcal{A}$ whose spectrum lies in the open right half plane. Then there exists a bounded derivation δ such that $e^\delta = \alpha$, and if $H^1(\mathcal{A}, \mathcal{A}) = 0$ then α is inner, and is implemented by an element e^a for some $a \in \mathcal{A}$.*

This may be rephrased as a perturbation result as follows.

7.2.2 Theorem. *Let $\mathcal{A}$ be a Banach algebra for which $H^1(\mathcal{A}, \mathcal{A}) = 0$. Then each automorphism α of $\mathcal{A}$ satisfying $\|\alpha - I\| < 1$ is an inner automorphism.*

Proof. The condition $\|\alpha - I\| < 1$ implies that the spectrum of α lies in the open right half plane, and the previous theorem may then be applied.

The cohomological condition in this theorem is essential. Without it, it is easy to construct counterexamples to the conclusion of Theorem 7.2.2.

7.2.3 Example. Let $\mathcal{A}_4 \subseteq \mathsf{M}_4$ be the algebra of matrices

$$\begin{pmatrix} * & 0 & 0 & 0 \\ 0 & * & 0 & 0 \\ * & * & * & 0 \\ * & * & 0 & * \end{pmatrix}$$

where $*$ denotes an arbitrary complex number. Let $\alpha_\varepsilon \colon \mathcal{A}_4 \to \mathcal{A}_4$ be the map which multiplies the (3,1) entry by $1 - \varepsilon$ and leaves the others unchanged. For $\varepsilon \neq 1$, α_ε is an automorphism of $\mathcal{A}_4$ and $\|\alpha_\varepsilon - I\| = |\varepsilon|$. No α_ε (for $\varepsilon \neq 0$) preserves the rank of matrices in $\mathcal{A}_4$ and so is not inner. Consequently no neighbourhood of I, however small, can consist entirely of inner automorphisms. The first cohomology group of this algebra is $\mathbb{C}$ [GHL].

Let $\mathcal{A}^{-1}$ denote the set of invertible elements in $\mathcal{A}$ and let $\mathrm{Aut}(\mathcal{A})$ denote the set of bounded automorphisms of $\mathcal{A}$. Both $\mathcal{A}^{-1}$ and $\mathrm{Aut}(\mathcal{A})$ are topological groups in the appropriate norm topologies, and there is a continuous homomorphism $\alpha \colon \mathcal{A}^{-1} \to \mathrm{Aut}(\mathcal{A})$ given by

$$\alpha_a(x) = axa^{-1}, \quad x \in \mathcal{A}, \quad a \in \mathcal{A}^{-1}.$$

The *principal component* of a topological group is the connected component which contains the identity. The following result is standard and may be found in [HewR].

7.2.4 Lemma. *Let G be a topological group and let W be an open connected neighbourhood of the identity. Then the principal component of G is the group generated by W.*

If G is the group $\mathcal{A}^{-1}$ of invertible elements in a Banach algebra, then we may take W to be $\{a \colon \|1 - a\| < 1\}$. The functional calculus shows that each element in W is an exponential e^b for some $b \in \mathcal{A}$, and it follows easily that the principal component consists of all finite products of exponentials.

7.2.5 Theorem. *Let $\mathcal{A}$ be a unital Banach algebra for which $H^1(\mathcal{A}, \mathcal{A}) = 0$. Then an automorphism ϕ is in the principal component of $\mathrm{Aut}\,(\mathcal{A})$ if* *and only if it has the form α_a for an element a in the principal component of $\mathcal{A}^{-1}$. If this is the case, then there exist $a_1, \ldots, a_n \in \mathcal{A}$ such that*

$$\phi(x) = e^{a_1} e^{a_2} \ldots e^{a_n} x e^{-a_n} \ldots e^{-a_2} e^{-a_1}, \quad x \in \mathcal{A}.$$

Proof. Let U and V be respectively the principal components of $\mathcal{A}^{-1}$ and $\mathrm{Aut}(\mathcal{A})$, let $W = \{\psi \in \mathrm{Aut}(\mathcal{A}): \|I - \psi\| < 1\}$, and let H be the subgroup of $\mathrm{Aut}(\mathcal{A})$ generated by $\{e^\delta: \delta$ is a derivation of $\mathcal{A}\}$. A typical element $e^{\delta_1} e^{\delta_2} \ldots e^{\delta_n}$ may be connected to the identity by the path

$$t \to e^{t\delta_1} e^{t\delta_2} \ldots e^{t\delta_n}, \quad 0 \le t \le 1,$$

and so H is a connected subgroup of $\mathrm{Aut}(\mathcal{A})$. By Lemma 7.2.4, each $\psi \in W$ has the form e^δ for some derivation δ on $\mathcal{A}$, and so the group H_1 generated by W is contained in H. The typical element of H may be expressed as

$$\psi_1^{\pm 1} \psi_2^{\pm 1} \ldots \psi_n^{\pm 1}, \quad \psi_i \in W,$$

and so H_1 is the union of open sets of the form

$$W \psi_2^{\pm 1} \ldots \psi_n^{\pm 1} \quad \text{or} \quad W^{-1} \psi_2^{\pm 1} \ldots \psi_n^{\pm 1}.$$

Consequently H_1 is an open subgroup of $\mathrm{Aut}(\mathcal{A})$. Its complement may be written as a union of cosets which implies that H_1 is also closed. Then the connectedness of H allows us to conclude that $H_1 = H$. Thus H is an open connected neighbourhood of the identity and so, by Lemma 7.2.4, $V = H$.

By hypothesis, each derivation δ is implemented by an element $a \in \mathcal{A}$, and so the automorphism e^δ is implemented by the element e^a. We have already shown that V consists of products of exponentials of derivations and so each $\phi \in V$ has the form

$$\phi(x) = e^{a_1} e^{a_2} \ldots e^{a_n} x e^{-a_n} \ldots e^{-a_2} e^{-a_1}, \quad x \in \mathcal{A}.$$

We have previously observed that U consists of finite products of exponentials, from which it is clear that α maps U onto V.

In the reverse direction, any automorphism ϕ of the above form can be connected to the identity by the path

$$\phi_t(x) = e^{ta_1} e^{ta_2} \ldots e^{ta_n} x e^{-ta_n} \ldots e^{-ta_2} e^{-ta_1}, \quad x \in \mathcal{A},$$

for $t \in [0, 1]$. Such an automorphism is then in the principal component.

7.3 An Implicit Function Theorem

Our aim is to present a result on the stability of products in the next section, and to accomplish this we will need an implicit function theorem for Banach space valued functions. These results are due independently to Johnson [J6] and Raeburn and Taylor [RaT]; we will follow the treatment in the latter paper with some minor changes.

The theorem concerns Fréchet differentiable functions on Banach spaces. Rather than digress for a discussion of differentiability, we have chosen to incorporate the derivatives into the hypotheses. When we apply this result to products, we will exhibit the appropriate derivatives explicitly.

If a linear map $g: \mathcal{X} \to \mathcal{Y}$ between Banach spaces has a closed range then it induces an isomorphism between $\mathcal{X}/\ker g$ and im g. We will refer to the norm of the inverse of this isomorphism as the *inversion constant* of g.

7.3.1 Theorem. *Let $\mathcal{X}, \mathcal{Y}$ and $\mathcal{Z}$ be Banach spaces and let $U \subseteq \mathcal{X}$ and $V \subseteq \mathcal{Y}$ be open convex subsets. Let $f: U \to V$, $k: V \to \mathcal{Z}$, $f': U \to \mathcal{L}^1(\mathcal{X}, \mathcal{Y})$ and $k': V \to \mathcal{L}^1(\mathcal{Y}, \mathcal{Z})$ be maps satisfying*

(a) im $f'(u) = \ker k'(f(u))$ for $u \in U$,

(b) $k'(f(u))$ has closed range for $u \in U$, and the inversion constants are uniformly bounded on U,

(c) the inversion constants for $f'(u)$ are uniformly bounded over U,

(d) there is a constant K such that if $u, u + x \in U$, $v, v + y \in V$ then

$$\text{(i)} \quad \|f(u + x) - f(u) - f'(u)x\| \le K\|x\|^2,$$

$$\text{(ii)} \quad \|k(v + y) - k(v) - k'(v)y\| \le K\|y\|^2,$$

(e) $k(f(u))$ is a constant function of $u \in U$.

If $u_0 \in U$ and $v_0 \in V$ satisfy $f(u_0) = v_0$, then there exist constants $\delta > 0$ and $C > 0$ such that for each $v \in V$ with $\|v_0 - v\| < \delta$ and $k(v) = k(v_0)$, there exists $u \in U$ with $\|u_0 - u\| \le C\|v_0 - v\|$ and $f(u) = v$.

Proof. From the first three hypotheses there is a constant M such that
(iii) if $y \in \ker k'(f(u))$ then there exists $x \in \mathcal{X}$ such that $f'(u)x = y$ and $\|x\| \le M\|y\|$,

(iv) if $z \in \text{im } k'(f(u))$ then there exists $y \in \mathcal{Y}$ such that $k'(f(u)y = z$ and $\|y\| \le M\|z\|$.

If $\delta > 0$, define constants r, C by

$$r = (MK + K(M + M^2 K)^2)\delta,$$
$$C = (1 - r)^{-1}(M + M^2 K).$$

Now fix δ sufficiently small that $\delta < 1$, $r < 1$ and the closed ball of radius $C\delta$ centred at u_0 is contained in U. Let $v \in V$ satisfy $k(v) = k(v_0)$ and $\|v - v_0\| < \delta$. We will obtain the solution to $f(u) = v$ as the limit of a sequence of approximations, the first of which is u_0.

By (iv) there exists $y_1 \in \mathcal{Y}$ satisfying

$$k'(v_0)y_1 = k'(v_0)(v - v_0),$$
$$\|y_1\| \le M\|k'(v_0)(v - v_0)\|.$$

Since $k(v_0) = k(v)$, it follows from (ii) that

$$\|k'(v_0)(v - v_0)\| = \|k(v) - k(v_0) - k'(v_0)(v - v_0)\| \le K\|v - v_0\|^2.$$

Thus

$$\|y_1\| \le MK\|v - v_0\|^2.$$

By construction $v - v_0 - y_1 \in \ker k'(v_0)$ and so, from (iii), there exists $x_1 \in \mathcal{X}$ such that

$$f'(u_0)x_1 = v - v_0 - y_1,$$
$$\|x_1\| \le M\|v - v_0 - y_1\|.$$

Then

$$\begin{aligned}
\|x_1\| &\le M\|v - v_0\| + M\|y_1\| \\
&\le M\|v - v_0\| + M^2K\|v - v_0\|^2 \\
&\le (M + M^2K\delta)\|v - v_0\| \\
&\le (M + M^2K)\|v - v_0\| \\
&\le C\delta.
\end{aligned}$$

Thus we may define u_1 to be $u_0 + x_1$ and $u_1 \in U$. Let $v_1 = f(u_1)$ and observe that

$$\begin{aligned}
\|v - v_1\| &= \|v - f(u_0 + x_1)\| \\
&= \|v - v_0 + f(u_0) - f'(u_0)x_1 + f'(u_0)x_1 - f(u_0 + x_1)\| \\
&\le \|v - v_0 - f'(u_0)x_1\| + \|f(u_0 + x_1) - f(u_0) - f'(u_0)x_1\| \\
&\le \|y_1\| + K\|x_1\|^2 \\
&\le \|y_1\| + KM^2\|v - v_0 - y_1\|^2 \\
&\le \|y_1\| + KM^2(\|v - v_0\| + \|y_1\|)^2 \\
&\le MK\|v - v_0\|^2 + KM^2(1 + MK)^2\|v - v_0\|^2 \\
&= (MK + K(M + M^2K)^2)\|v - v_0\|^2 \\
&\le r\|v - v_0\|.
\end{aligned}$$

By (e),

$$k(v_1) = k(f(u_1)) = k(f(u_0)) = k(v_0) = k(v).$$

Now suppose that we have constructed $\{u_i\}_{i=0}^n \in U$, $\{v_i\}_{i=0}^n \in V$ such that

$$f(u_i) = v_i,$$
$$k(v_i) = k(v),$$
$$\|u_{i+1} - u_i\| \le (M + M^2K)r^i\|v - v_0\|,$$
$$\|v - v_i\| \le r^i\|v - v_0\|,$$

for $0 \le i \le n$. Then repeat the above argument with u_0 and v_0 replaced by u_n and v_n to obtain u_{n+1}, v_{n+1} satisfying

$$v_{n+1} = f(u_{n+1}),$$
$$k(v_{n+1}) = k(v),$$
$$\|v - v_{n+1}\| \le r\|v - v_n\| \le r^{n+1}\|v - v_0\|,$$
$$\|u_{n+1} - u_n\| \le (M + M^2K)\|v - v_n\| \le (M + M^2K)r^n\|v - v_0\|.$$

We are guaranteed that $u_{n+1} \in U$ since

$$
\begin{aligned}
\|u_{n+1} - u_0\| = \left\|\sum_{i=0}^n (u_{i+1} - u_i)\right\| \\
\le \sum_{i=0}^n (M + M^2K)r^i\|v - v_0\| \\
\le (1 - r)^{-1}(M + M^2K)\|v - v_0\| \\
= C\|v - v_0\| \\
\le C\delta.
\end{aligned}
$$

The sequence $\{u_n\}_{n=0}^\infty$ is Cauchy, so converges to $u \in \mathcal{X}$. Then $\|u - u_0\| \le C\delta$ so $u \in U$. It is clear from (d) that f is continuous at u, so

$$f(u) = \lim_{n \to \infty} f(u_n) = \lim_{n \to \infty} v_n = v.$$

In addition,

$$\|u - u_0\| = \lim_{n \to \infty} \|u_{n+1} - u_0\| \le C\|v - v_0\|.$$

7.4 Stability

The standard product $(x, y) \to xy$ on a von Neumann algebra $\mathcal{M}$ is a continuous bilinear map $m_0 \colon \mathcal{M} \times \mathcal{M} \to \mathcal{M}$ which satisfies

$$m_0(x, m_0(y, z)) = m_0(m_0(x, y), z)$$

(associativity). Any continuous bilinear map $m: \mathcal{M} \times \mathcal{M} \to \mathcal{M}$ which obeys this equation is called an associative product. These arise naturally: given an invertible map $\phi: \mathcal{M} \to \mathcal{M}$ we may define

$$m(x, y) = \phi^{-1}(\phi(x)\phi(y)), \quad x, y \in \mathcal{M}.$$

Continuity of m is a consequence of the simple estimate

$$\|m(x, y)\| \leq \|\phi^{-1}\| \, \|\phi\|^2 \|x\| \, \|y\|.$$

The main result of this section will show that in certain circumstances every associative product which is close to m_0 in the norm of $\mathcal{L}^2(\mathcal{M}, \mathcal{M})$ has this form, where ϕ may be chosen close to the identity. We formulate the result for von Neumann algebras but the proof is valid for general Banach algebras.

The determination of cohomology groups in the previous two chapters shows that the hypotheses are satisfied by all von Neumann algebras of types I, II_∞ or III, all hyperfinite von Neumann algebras, and all type II_1 von Neumann algebras which have Cartan subalgebras or are stable.

7.4.1 Theorem. *Let $\mathcal{M}$ be a von Neumann algebra such that $H^2(\mathcal{M}, \mathcal{M}) = H^3(\mathcal{M}, \mathcal{M}) = 0$. Then there exist constants $C, \delta > 0$ such that if m is an associative product on $\mathcal{M}$ and $\|m - m_0\| < \delta$ then there exists an invertible map $\phi: \mathcal{M} \to \mathcal{M}$ with $\|\phi - I\| \leq C\|m - m_0\|$ and*

$$m(x, y) = \phi^{-1}(\phi(x)\phi(y)), \quad x, y \in \mathcal{M}.$$

Proof. In the notation of Theorem 7.3.1, let $\mathcal{X} = \mathcal{L}^1(\mathcal{M}, \mathcal{M})$, $\mathcal{Y} = \mathcal{L}^2(\mathcal{M}, \mathcal{M})$, $\mathcal{Z} = \mathcal{L}^3(\mathcal{M}, \mathcal{M})$. Define $f: \mathcal{L}^1(\mathcal{M}, \mathcal{M})^{-1} \to \mathcal{L}^2(\mathcal{M}, \mathcal{M})$ by

$$f(\phi)(x, y) = \phi^{-1}(\phi(x)\phi(y)), \quad x, y \in \mathcal{M}, \phi \in \mathcal{L}^1(\mathcal{M}, \mathcal{M})^{-1}.$$

Then define $k: \mathcal{L}^2(\mathcal{M}, \mathcal{M}) \to \mathcal{L}^3(\mathcal{M}, \mathcal{M})$ by

$$k(\alpha)(x, y, z) = \alpha(x, \alpha(y, z)) - \alpha(\alpha(x, y), z), \quad x, y, z \in \mathcal{M}, \ \alpha \in \mathcal{L}^2(\mathcal{M}, \mathcal{M}).$$

The kernel of k is the set of associative products and so $k(f(\phi)) = 0$ for $\phi \in \mathcal{L}^1(\mathcal{M}, \mathcal{M})^{-1}$.

If $\phi \in \mathcal{L}^1(\mathcal{M}, \mathcal{M})^{-1}$ and $\psi \in \mathcal{L}^1(\mathcal{M}, \mathcal{M})$ has small norm then $(\phi + \psi)^{-1}$ exists and is expressible as the sum $\phi^{-1}\left(\sum_{n=0}^{\infty}(-1)^n(\psi\phi^{-1})^n\right)$. Thus

$$(f(\phi + \psi) - f(\phi))(x, y) = (\phi + \psi)^{-1}((\phi(x) + \psi(x))(\phi(y) + \psi(y)))$$
$$- \phi^{-1}(\phi(x)\phi(y))$$

$$= \phi^{-1}\left(\sum_{n=0}^{\infty}(-1)^n(\psi\phi^{-1})^n\right)(\phi(x)\phi(y) + \psi(x)\phi(y)$$

$$+ \phi(x)\psi(y) + \psi(x)\psi(y)) - \phi^{-1}(\phi(x)\phi(y))$$

$$= \phi^{-1}[\psi(x)\phi(y) - \psi(\phi^{-1}(\phi(x)\phi(y))) + \phi(x)\psi(y)]$$

$$+ \text{higher-order terms in } \psi.$$

Then define

$$(f'(\phi)\psi)(x,y) = \phi^{-1}[\phi(x)\psi(y) - \psi(\phi^{-1}(\phi(x)\phi(y)) + \psi(x)\phi(y)].$$

A simple estimate in the series above shows that, for $\phi, \phi+\psi$ in a sufficiently small ball centred at I,

$$\|f(\phi + \psi) - f(\phi) - f'(\phi)\psi\|$$

will be dominated by a multiple of $\|\psi\|^2$.

Turning to k, we have

$$\begin{aligned}
k(\alpha + \beta)(x,y,z) &= (\alpha + \beta)(x, \alpha(y,z) + \beta(y,z)) \\
&\quad - (\alpha + \beta)(\alpha(x,y) + \beta(x,y), z) \\
&= k(\alpha)(x,y,z) + [\alpha(x, \beta(y,z)) - \beta(\alpha(x,y),z) \\
&\quad + \beta(x, \alpha(y,z)) - \alpha(\beta(x,y),z)] \\
&\quad + \beta(x, \beta(y,z)) - \beta(\beta(x,y),z).
\end{aligned}$$

We may then define $(k'(\alpha)\beta)(x,y,z)$ to be the element inside $[\cdot]$ and we immediately obtain

$$\|k(\alpha + \beta) - k(\alpha) - k'(\alpha)\beta\| \leq 2\|\beta\|^2.$$

Thus we have verified the last two hypotheses of Theorem 7.3.1 on a sufficiently small neighbourhood of I.

Now

$$\begin{aligned}
f'(I)\psi(x,y) &= x\psi(y) - \psi(xy) + \psi(x)y \\
&= \partial\psi(x,y),
\end{aligned}$$

while

$$\begin{aligned}
(k'(f(I))\beta)(x,y,z) &= (k'(m_0)\beta)(x,y,z) \\
&= x\beta(y,z) - \beta(xy,z) + \beta(x,yz) - \beta(x,y)z \\
&= \partial\beta(x,y,z).
\end{aligned}$$

The image of $f'(I)$ is the space of 2-coboundaries while the kernel of $k'(m_0)$ is the space of 2-cocycles; by hypothesis these are equal. More generally

$$\begin{aligned}
k'(f(\phi))\beta)(x,y,z) &= \phi^{-1}(\phi(x)\beta(y,z)) - \beta(\phi^{-1}(\phi(x)\phi(y)),z) \\
&\quad + \beta(x, \phi^{-1}(\phi(y)\phi(z))) - \phi^{-1}\phi(\beta(x,y)\phi(z)) \\
&= \phi^{-1}(\partial\widetilde{\beta}(\phi(x), \phi(y), \phi(z)))
\end{aligned}$$

where $\widetilde{\beta}$ is defined by

$$\widetilde{\beta}(x,y) = \phi(\beta(\phi^{-1}(x), \phi^{-1}(y))).$$

Thus $\beta \in \ker k'(f(\phi))$ if and only if $\widetilde{\beta}$ is a 2-cocycle, which is equivalent to writing

$$\widetilde{\beta}(x, y) = x\psi(y) - \psi(xy) + \psi(x)y.$$

Thus

$$\begin{aligned}
\beta(x, y) &= \phi^{-1}\widetilde{\beta}(\phi(x), \phi(y)) \\
&= \phi^{-1}(\phi(x)\psi(\phi(y)) - \phi^{-1}\psi(\phi(x)\phi(y)) - \phi^{-1}(\psi(\phi(x))\phi(y)) \\
&= (f'(\phi)\widetilde{\psi})(x, y)
\end{aligned}$$

where

$$\widetilde{\psi}(x) = \psi(\phi(x)).$$

Thus im $f'(\phi) = \ker k'(f(\phi))$, verifying the first hypothesis of Theorem 7.3.1.

The range of $k'(m_0)$ is the space of 3-coboundaries on $\mathcal{M}$ which, by hypothesis, equals the space of 3-cocycles. Thus the range of $k'(m_0)$ is closed.

Since

$$(k'(f(\phi))\beta)(x, y, z) = \phi^{-1}\partial\widetilde{\beta}(\phi(x), \phi(y), \phi(z))$$

and

$$(f'(\phi)\widetilde{\psi})(x, y) = \phi^{-1}\partial\psi(\phi(x), \phi(y)),$$

the inversion constants of these maps will be bounded on any neighbourhood of I on which $\|\phi\|$ and $\|\phi^{-1}\|$ are uniformly bounded.

We have now verified all the hypotheses of Theorem 7.3.1, since

$$k(m) = k(m_0) = 0$$

for any associative product. The existence of ϕ satisfying

$$m(x, y) = \phi^{-1}(\phi(x)\phi(y))$$

is now assured, and the norm estimates follow from Theorem 7.3.1.

In Theorem 7.4.1 the product might satisfy

$$m(x, y)^* = m(y^*, x^*).$$

In this case the map $\phi\colon \mathcal{M} \to \mathcal{M}$ which implements m may be chosen to be self-adjoint. The appropriate modifications in the proof are:

$$\begin{aligned}
\mathcal{X} &= \{\phi \in \mathcal{L}^1(\mathcal{M}, \mathcal{M})\colon \phi(x)^* = \phi(x^*)\}, \\
\mathcal{Y} &= \{\alpha \in \mathcal{L}^2(\mathcal{M}, \mathcal{M})\colon \alpha(x, y)^* = \alpha(y^*, x^*)\}, \\
\mathcal{Z} &= \{\psi \in \mathcal{L}^3(\mathcal{M}, \mathcal{M})\colon \psi(x, y, z)^* = -\psi(z^*, y^*, x^*)\}.
\end{aligned}$$

We also note that instead of the hypothesis that $H^3(\mathcal{M}, \mathcal{M}) = 0$ we could assume the weaker condition that the space of 3-coboundaries is closed.

We conclude the chapter with two results, the first of which shows that certain linear isomorphisms are close to algebraic isomorphisms.

7.4.2 Corollary. *Let $M \subseteq B(H)$ be a von Neumann algebra such that $H^2(M, M) = H^3(M, M) = 0$. Then there exist constants $C, \delta > 0$ such that if B is a closed subalgebra of $B(H)$ and $\lambda\colon M \to B$ is a linear isomorphism satisfying $\|\lambda - I\| < \delta$, then there is an algebraic isomorphism $\phi\colon M \to B$ satisfying $\|\phi - \lambda\| \leq C\|\lambda - I\|$. If in addition B is a C^*-algebra, then ϕ may be chosen to be a $*$-isomorphism.*

Proof. Define an associative product on M by

$$m(x, y) = \lambda^{-1}(\lambda(x)\lambda(y)), \quad x, y \in M.$$

By Theorem 7.4.1, there is a linear invertible map $\psi\colon M \to M$ such that

$$m(x, y) = \psi^{-1}(\psi(x)\psi(y)), \quad x, y \in M.$$

Then $\phi = \lambda\psi^{-1}\colon M \to B$ is an algebraic isomorphism close to λ.

If B is closed under the adjoint operation then we may assume that λ is self-adjoint by replacing it, if necessary, by $x \to (\lambda(x) + \lambda(x^*)^*)/2$. Then m is a self-adjoint product and so ψ above may be chosen to be self-adjoint. It follows that $\phi = \lambda\psi^{-1}$ is a $*$-isomorphism.

We give one application of this result to the topic of close von Neumann algebras. Given von Neumann algebras M, N contained in $B(H)$, the distance $d(M, N)$ between them is defined by

$$d(M, N) = \sup\{\|m - N_1\|,\ \|n - M_1\|\colon m \in M_1, n \in N_1\},$$

where M_1 and N_1 are the unit balls of M and N respectively and

$$\|m - N_1\| = \inf\{\|m - n\|\colon n \in N_1\}$$

with a similar definition for $\|n - M_1\|$. Thus closeness means that elements in each unit ball are close to elements in the other unit ball. Kadison and Kastler [KK] posed the problem of determining whether close von Neumann algebras (or C^*-algebras) are isomorphic. We can now answer this if one of the algebras is injective, which of course means that it is hyperfinite.

7.4.3 Corollary. *Let M be an injective subalgebra of $B(H)$. Then there exist constants $\delta > 0$, $C > 0$ such that if $N \subseteq B(H)$ is another operator algebra and $d(M, N) < \delta$ then there exists a $*$-isomorphism $\phi\colon M \to N$ with $\|\phi - I\| \leq Cd(M, N)$.*

Proof. Since M is injective there exists a norm one projection $P\colon B(H) \to M$. Let $\mu\colon N \to M$ denote its restriction to N. Suppose that $d(M, N) < \varepsilon$

with $\varepsilon < 1/2$. If $x \in \mathcal{N}$ with $\|x\| = 1$ then there exists $y \in \mathcal{M}$ such that $\|y\| \leq 1$ and $\|x - y\| < \varepsilon$. Then $Py = y$, so

$$\|\mu(x) - x\| = \|P(x - y) - (x - y)\| \leq 2\varepsilon,$$

and for a general element $x \in \mathcal{N}$,

$$\|\mu(x) - x\| \leq 2\varepsilon \|x\|.$$

An element $x \in \ker \mu$ must satisfy

$$\|x\| \leq 2\varepsilon \|x\|$$

which forces $x = 0$, and so μ is injective. We also have, for $x \in \mathcal{N}$,

$$\|x\| - \|\mu(x)\| \leq \|x - \mu(x)\| \leq 2\varepsilon \|x\|$$

from which it follows that

$$(1 - 2\varepsilon)\|x\| \leq \|\mu(x)\|.$$

Then any sequence $\{x_n\}_{n=1}^{\infty}$ for which $\{\mu(x_n)\}_{n=1}^{\infty}$ is Cauchy must itself be Cauchy. Thus, if $\lim_{n \to \infty} \mu(x_n) = y \in \mathcal{M}$, then $\{x_n\}_{n=1}^{\infty}$ has a limit $x \in \mathcal{N}$ and $\mu(x) = y$. This shows that the range of μ (which we denote by Ran μ) is closed.

Suppose now that Ran μ is strictly contained in $\mathcal{M}$. Then for each $y \in \mathcal{M}$, $\|y\| = 1$, we may find $x \in \mathcal{N}$ such that

$$\|y - x\| \leq \varepsilon$$

and so

$$\|y - \mu(x)\| \leq \varepsilon.$$

Thus the quotient map of $\mathcal{M}$ onto $\mathcal{M}/\text{Ran } \mu$ has norm at most $\varepsilon < 1$, a contradiction. We conclude that $\mu \colon \mathcal{N} \to \mathcal{M}$ is surjective and so is invertible. Denote its inverse by $\lambda \colon \mathcal{M} \to \mathcal{N}$. Observe that the hyperfiniteness of $\mathcal{M}$ gives $H^2(\mathcal{M}, \mathcal{M}) = H^3(\mathcal{M}, \mathcal{M}) = 0$ (Corollary 3.4.6), and so the existence of the desired isomorphism is a consequence of the previous corollary applied to λ.

7.5 Notes and Remarks

The preliminary work in Section 7.2 is taken from [BD2], but is originally due to [ZeM]. The connections between cohomology and the stability of products in von Neumann algebras (and more generally in Banach algebras) were investigated simultaneously and independently by Johnson [J6] and Raeburn and Taylor [RaT]. Our approach is essentially from the latter paper, with some minor modifications.

Several authors have considered the perturbation theory of operator algebras, and the reader is referred to [Ch1, Ch2, Ch6, J6, KK, L, PhR1, PhR2, RaT, Ta] for more information on the topic. In particular, the survey article by Christensen [Ch6] contains much of what is currently known.

8

Appendix

8.1 Introduction

This appendix contains a section on bounded group cohomology and one on its relation with the ℓ^1-group algebra cohomology. Though there is currently no link between bounded group cohomology and that of the reduced group or von Neumann group algebras it is an obvious question to ask if the subjects are related. Given that bounded group cohomology is a topic unknown to most operator algebraists, it seemed worth introducing it in Section 8.2 and linking it with Hochschild cohomology in Section 8.3. There is a list of problems in Section 8.4.

8.2 Bounded Group Cohomology

8.2.1 Remarks. Bounded group cohomology was related to corresponding geometrical and topological ideas for manifolds by Gromov in (1982) [Grom] following work of Hirsch and Thurston (1975) [HiT]. Earlier Johnson (1972) [J3] had used bounded cohomology of groups to show that $H^2(\ell^1(G), \ell^1(G)) \neq 0$ for G the free group on two generators. Bounded group cohomology is defined and a few of its properties are given in these notes. The theory is only developed as far as its current relevance to the Hochschild cohomology of Banach algebras warrants. For further details of the theory see the paper by Gromov [Grom] (beware there are errors), the survey by Ivanov [Iv1] and the paper by Grigorchuk [Gri2]. The authors are indebted to Professor Grigorchuk for the preprint [Gri2], which is recommended reading.

An elementary concrete approach is taken to the bounded cohomology of groups analogous to the Hochschild cohomology discussion.

8.2.2 Definition. Throughout this section G will denote a discrete group. The groups the reader should keep in mind are finitely generated non-abelian infinite groups like the free group $\mathbf{F}_n$ on n-generators ($n \geq 2$) or a general hyperbolic group. Let $\mathcal{X}$ be a Banach space, which is usually $\mathbf{R}$ on $\mathbf{C}$ below.

Let $C_b^n(G, \mathcal{X})$ be the space of bounded functions from G into the Banach space $\mathcal{X}$, and let $C_b^n(G, \mathcal{X})$ have the $\|\cdot\|_\infty$-norm. The differential $\delta = \delta^n$ is defined by

$$(\delta f)(x_1, \ldots, x_{n+1}) = f(x_2, \ldots, x_{n+1})$$
$$+ \sum_{j=1}^{n} (-1)^j f(x_1, \ldots, x_{j-1}, x_j x_{j+1}, \ldots, x_{n+1})$$
$$+ (-1)^{n+1} f(x_1, \ldots, x_n)$$

for all $f \in C_b^n(G, \mathcal{X})$ and $x_1, \ldots, x_{n+1}$ in G. Note that δ is a continuous linear operator with $\|\delta^n\| \leq n + 2$. A tedious elementary calculation shows that $\delta \cdot \delta = 0$; the group is assumed to have trivial action on $\mathcal{X}$. The bounded cohomology group of G with values in $\mathcal{X}$ (with trivial group action) is $H_b^n(G, \mathcal{X})$ equal to

$$\frac{\mathrm{Ker}(\delta\colon\ C_b^n(G, \mathcal{X}) \to C_b^{n+1}(G, \mathcal{X}))}{\mathrm{Im}(\delta\colon\ C_b^{n-1}(G, \mathcal{X}) \to C_b^n(G, \mathcal{X}))};$$

it is the cohomology of the cochain complex $(C_b^*(G, \mathcal{X}), \delta)$. If $\mathcal{X} = \mathbf{R}$, we write $H_b^n(G)$ instead of $H_b^n(G, \mathbf{R})$.

One of the reasons this theory is interesting is the following theorem due to Gromov and Brooks [Grom], [Br] (see also [Iv1] for another proof). This result links the bounded cohomology of groups with the bounded cohomology of manifolds.

8.2.3 Theorem. *Let Y be a connected manifold. Then the bounded cohomology $H_b^n(Y)$ of the manifold is isomorphic to the bounded cohomology $H_b^n(\pi_1(Y))$ of the fundamental group $\pi_1(Y)$ of Y. This isomorphism preserves the seminorms on the two spaces.*

8.2.4 Remarks.

(a) Observe that

$$H_b^n(G, \mathbb{C}) = H_b^n(G) + i H_b^n(G)$$

just by splitting functions into their real and imaginary parts. Care is required with the seminorms.

(b) Note that $H_b^1(G) = 0$ for all groups G, because a function f is in $\mathrm{Ker}(\delta\colon C_b^1(G) \to C_b^2(G))$ if and only if f is constant.

(c) Though it is unknown if $H_b^n(G)$ is a Banach space for all n, Ivanov [Iv2] proved that $H_b^2(G)$ is a Banach space.

8.2.5 Theorem. *For all discrete groups G, $H_b^2(G)$ is a Banach space in the quotient seminorm.*

Proof. It is sufficient to construct a bounded linear map γ from $C_b^2(G)$ onto $C_b^1(G)$ such that $\gamma\delta = I_1 = $ identity on $C_b^1(G)$, where δ maps $C_b^1(G)$ into $C_b^2(G)$ in this proof. Let I_2 be the identity on $C_b^2(G)$, and let $Q = I_2 - \delta\gamma$. Then

$$Q\delta = \delta - \delta(\gamma\delta) = \delta - \delta = 0$$

so $\mathrm{Im}\ \delta \subseteq \mathrm{Ker}\ Q$. If $f \in \mathrm{Ker}\ Q$, then $f = \delta\gamma f \in \mathrm{Im}\ \delta$. Hence $\mathrm{Im}\ \delta = \mathrm{Ker}\ Q$ is closed, and $H_b^2(G)$ is a Banach space. It remains to construct γ. Let LIM denote a Banach limit on $\mathbb{N}$ so that $LIM1 = 1$, $LIM\alpha_{n+1} = LIM\alpha_n$ and

$\liminf \alpha_n \leq LIM\alpha_n \leq \limsup \alpha_n$ for all bounded sequences (α_n). Adding the equation

$$(\delta f)(x^j, x) = f(x^j) - f(x^{j+1}) + f(x)$$

for $1 \leq j \leq n - 1$ implies that

$$\sum_{j=1}^{n-1}(\delta f)(x^j, x) = nf(x) - f(x^n)$$

so that

(1)
$$f(x) = n^{-1}f(x^n) + n^{-1}\sum_{j=1}^{n-1}(\delta f)(x^j, x)$$

for all x in G. Now define $\gamma \colon \mathcal{C}_b^2(G) \to \mathcal{C}_b^1(G)$ by

$$(\gamma F)(x) = LIM n^{-1}\sum_{j=1}^{n-1} F(x^j, x)$$

for all x in G. Note that γ is a continuous linear map with $\|\gamma\| = 1$. Applying LIM to equation (1) gives $f(x) = (\gamma \delta f)(x)$ for all x in G as required.

In the Hochschild cohomology of von Neumann algebras an averaging argument implied that the cohomology was isomorphic to equivariant cohomology with respect to a hyperfinite (= von Neumann amenable) subalgebra. The following result is a bounded group version of this due to Ivanov [Iv1] (1985; see the paper for a proof).

8.2.6 Theorem. *Let K be an amenable normal subgroup of G. The map q between the bounded complexes of G/K and G,*

$$q \colon C_b^n(G/K) \to C_b^n((G),$$

defined by

$$(qf)(x_1, \ldots, x_n) = f(x_1 K, \ldots, x_n K)$$

induces an isomorphism from $H_b^n(G/K)$ onto $H_b^n(G)$ that preserves the seminorms.

8.2.7 Remarks.
(a) From the above theorem one immediately deduces that if G is amenable, then $H_b^n(G) = 0$ for all n.
(b) The Gromov–Brooks theorem implies there are many discrete groups for which $H_b^2(G) \neq 0$. However from the point of view of the Hochschild cohomology of Banach algebras discussed in Section 8.3 it is desirable to have

some concrete bounded cocycles. The first were obtained by Johnson [J3] and in a more general form by Brooks [Br]. The latter are combinatorially elementary to define and handle so are discussed briefly below. However the natural bounded cocycles in $H_b^2(\mathsf{F}_n)$ are those of Faiziev [Fa], which are similar to Brooks' but more complicated. Faiziev's cocycles have the advantage that they are linearly independent whereas Brooks' ones are not (see [Gri2] for a discussion).

8.2.8 Definition. Let F_n be the free group on n (≥ 2) generators $g_1, g_2, \ldots,$ g_n with $n = \infty = \aleph_0$ included. If $x \in \mathsf{F}_n$ is written in reduced form as

$$x = g_{i_1}^{\varepsilon_1} \cdots g_{i_m}^{\varepsilon_m},$$

where $\varepsilon_j = \pm 1$ and there is no cancellation let the length $|x|$ of x be m. Note $|e| = 0$. If w ($\neq e$) is in F_n, let $\#w$ *in* x denote the number of times the word w occurs in the word x including repetitions in overlaps. For example, if $w = x_1^2$ and $x = x_1^4 x_2 x_1^{-2}$, then $\#w$ in x is 3. Note that

$$\#w^{-1} \text{ in } x = \#w \text{ in } x^{-1}.$$

The Brooks cocycle is now defined by

$$\beta_w(x) = \#w \text{ in } x - \#w \text{ in } x^{-1}.$$

The definition of the Faiziev cocycles, or pseudocharacters, is a little more intricate. If $x \in \mathsf{F}_n$, let $x = t x_c t^{-1}$, where x_c is the shortest word in the conjugacy class of x so x_c is cyclically reduced; i.e. the first symbol in x_c is not the inverse of the last. Let $\overline{x}_c$ be x_c thought of as a cyclic word, i.e. written around a circle, and let the Faiziev pseudocharacter be

$$\rho_w(x) = \#w \text{ in } \overline{x}_c - \#w^{-1} \text{ in } \overline{w}_c.$$

For example if $w = g_1 g_2^{-1}$, and $x = g_2^{-2} g_1 g_2^{-1} g_1 g_2$ then

$$t = g_2^{-1} \text{ and } x_c = g_2^{-1} g_1 g_2^{-1} g_1$$

so $\rho_w(x) = 2$. See [Fa] and [Gri2] for details. The Brooks cocycles are simple enough to include all the details here.

8.2.9 Lemma. *Let* $w \in \mathsf{F}_n$ *with* $w \neq e$. *Then*
(a) $\#w$ *in* $x \leq |x| - |w| - 1$;
(b) *if* $x, y \in \mathsf{F}_n$ *with no cancellation in* xy, *i.e.* $|xy| = |x| + |y|$, *then*

$$|\beta_w(xy) - \beta_w(x) - \beta_w(y)| \leq |w| - 1;$$

(c) $|\delta\beta_w(x,y)| \leq 3(|w|-1)$.

Proof. (a) The inequality follows from the pigeon hole principle or by seeing that an upper bound may be obtained by putting w in the first $|w|$ places in x and shifting one place to the right at a time until it is in the last $|w|$ places.

Note $-\#w^{-1}$ in $x \leq \beta_w(x) \leq \#w$ in x.

(b) Now

$$\#w \text{ in } xy = \#w \text{ in } x + \#w \text{ in } y + \#w \text{ in } uv,$$

where u are the last $|w|-1$ symbols in x and v are the first $|w|-1$ symbols in y.

Further

(1)
$$\#w \text{ in } uv \leq |uv| - |w| + 1$$
$$\leq 2|w| - 2 - |w| + 1$$
$$= |w| - 1$$

by (a). Since $|w| = |w^{-1}|$,

$$|\beta_w(xy) - \beta_w(x) - \beta_w(y)| = |\beta_w(uv)|$$
$$\leq |w| - 1$$

by (1).

Note that both parts of the above lemma are best possible; choose integers $1 \leq k < m, n$, and let $w = g_1^k$, $x = g_1^m$, $y = g_1^n$.

(c) Let $x = x_1 u$ and $y = u^{-1}y_1$ with $|x_1 y_1| = |x_1| + |y_1|$; thus u is the part of xy that cancels off. Using $\beta_w(u^{-1}) = -\beta_w(u)$ yields

$$|\delta\beta_w(x,y)| = |\beta_w(x_1 u) - \beta_w(x_1 y_1) + \beta_w(u^{-1}y_1)|$$
$$\leq |\beta_w(x_1 u) - \beta_w(x_1) - \beta_w(u)|$$
$$+ |\beta_w(x_1 y_1) - \beta_w(x_1) - \beta_w(y_1)|$$
$$+ |\beta_w(u^{-1}y_1) - \beta_w(u^{-1}) - \beta_w(y_1)|$$
$$\leq 3(|w| - 1)$$

by (b) as required.

8.2.10 Theorem. *For all $n \geq 2$,*

$$H^2(\mathsf{F}_n, \mathbf{R}) \neq 0.$$

Proof. Let $w = g_1 g_2$ and suppose that $\partial \beta_w = \partial \phi$ for some $\phi \in C_b^1(G, \mathbf{R})$. Let $\psi = \phi - \beta_w$ and let

$$V = \{w g_k^{\pm 1} w^{-1} \colon g_k \text{ generator}\}.$$

Then $\psi(xy) = \psi(x) + \psi(y)$ for all $x, y \in \mathbf{F}_n$. If $k \neq 2$ and $j \in \mathbf{Z}$, then

$$\#w \text{ in } (w g_k w^{-1})^j = \#w \text{ in } g_1 g_2 g_k^j g_2^{-1} g_1^{-1}$$
$$= \begin{cases} 0 & \text{for} \quad j = 0 \\ 1 & \text{for} \quad j \neq 0. \end{cases}$$

If $k = 2$ and $j \in \mathbf{Z}$, then

$$\#w \text{ in } (w g_k w^{-1})^j = \#w \text{ in } g_1 g_2^j g_1^{-1}$$
$$= \begin{cases} 0 & \text{for} \quad j \leq 0 \\ 1 & \text{for} \quad j \geq 1. \end{cases}$$

If $x \in V$ and $j \in \mathbf{Z}$, it follows that

$$\beta_w(x^j) \in \{-1, 0, 1\}$$

so that $|\beta_w(x^j)| \leq 1$ and $|\phi(x^j)| \leq \|\phi\|_\infty$. Then

$$\{|\psi(x^j)| = j|\psi(x)| \colon j \in \mathbf{N}\}$$

is bounded by $1 + \|\phi\|_\infty$, so $\psi(x) = 0$. Thus ψ vanishes on $\mathbf{F}_n$ since this group is generated by V. We conclude that β_w is bounded on $\mathbf{F}_n$, which contradicts the relations

$$\beta_w(w^j) = j, \qquad j \geq 1.$$

8.2.11 Remark. Note that Brooks [Br, 3.a, p. 58] states that the cocycles $\partial \beta_w$ are linearly independent in $H_b^2(\mathbf{F}_2)$ for $w \in \mathbf{F}_2$ with $|w| \geq 2$. This is incorrect (see [Gri2]) because the element $\delta \phi$ is zero on $H_b^2(G)$, where

$$\phi = \beta_{g_1 g_2} + \beta_{g_1^{-1} g_2} + \beta_{g_1 g_2^{-1}} + \beta_{g_1^{-1} g_2^{-1}}.$$

The function ϕ is bounded, since in $\mathbf{F}_2$

$$\phi(x) = \begin{cases} 1 & \text{if } x \text{ begins with } g_1^{\pm 1} \text{ and ends with } g_2^{\pm 1} \\ -1 & \text{if } x \text{ begins with } g_2^{\pm 1} \text{ and ends with } g_1^{\pm 1} \\ 0 & \text{otherwise}; \end{cases}$$

so ϕ is bounded.

See Grigorchuk's paper [Gri2] for the structure of several of the Banach spaces $H_b^2(G)$, for example, $G = \mathsf{F}_n$ ($n \geq 2$).

8.3 Bounded Group Cohomology and Hochschild Cohomology

Let tr denote the normalized trace on the convolution Banach algebra $\ell^1(G)$; so

$$tr\left(\sum_{x \in G} \alpha_x x\right) = \alpha_e.$$

The following theorem is essentially in [J3]; Johnson considers (1) only for $n = 2$, but the ideas are there.

8.3.1 Theorem. *Let G be a discrete group.*
(1) The linear map $P\colon C_b^n(G, \mathbb{C}) \to \mathcal{L}^n(\ell^1(G), \ell^\infty(G))$ defined by

$$P(\phi)(x_1, \ldots, x_n)(y) = \phi(x_1, \ldots, x_n)tr(x_1 \ldots x_n y)$$

for all $x_1, \ldots, x_n, y \in G$, is continuous, $\|P\| = 1$ and $\partial P = P\delta$. Further P induces an embedding of $H_b^n(G, \mathbb{C})$ into $H^n(\ell^1(G), \ell^\infty(G))$ as a direct summand.
(2) The linear map $Q\colon C_b^n(G, \mathbb{C}) \to \mathcal{L}^n(\ell^1(G), \ell^1(G))$ defined by

$$(Q\phi)(x_1, \ldots, x_n) = \phi(x_1, \ldots, x_n)x_1 \ldots x_n$$

for all $x_1, \ldots, x_n$ is continuous, $\|Q\| = 1$ and $\partial Q = Q\delta$. Further Q induces an embedding of $H_b^n(G, \mathbb{C})$ into $H^n(\ell^1(G), \ell^1(G))$ as a direct summand.

Proof. (1) Clearly P is a continuous linear operator from $C_b^n(G, \mathbb{C})$ into $\mathcal{L}^n(\ell^1(G)), \ell^\infty(G))$ with $\|P\| = 1$. If $x_1, \ldots, x_{n+1}$ are in $G \subseteq \ell^1(G)$, y is in $G \subseteq \ell^\infty(G)$ and ϕ is in $C_b^n(G, \mathbb{C})$, then

$$\partial(P\phi)(x_1, \ldots, x_{n+1})(y)$$
$$= x_1(P\phi)(x_2, \ldots, x_{n+1})(y)$$
$$\quad + \sum_{j=1}^n (-1)^j P\phi(x_1, \ldots, x_j x_{j+1}, \ldots, x_{n+1})(y)$$
$$\quad + (-1)^{n+1} P\phi(x_1, \ldots, x_n)x_{n+1}(y)$$
$$= \phi(x_2, \ldots, x_{n+1})tr(x_2 \ldots x_{n+1} y x_1)$$
$$\quad + \sum_{j=1}^n (-1)^j \phi(x_1, \ldots, x_j x_{j+1}, \ldots, x_{n+1})tr(x_1 \ldots x_{n+1} y)$$
$$\quad + (-1)^{n+1} \phi(x_1, \ldots, x_n)tr(x_1 \ldots x_{n+1} y)$$
$$= (\delta\phi)(x_1, \ldots, x_{n+1})tr(x_1 \ldots x_{n+1} y)$$

by the tracial property. Hence P induces a map from $H_b^n(G)$ into

$$H^n(\ell^1(G), \ell^\infty(G)).$$

Let $p\colon \mathcal{L}^n(\ell^1(G), \ell^\infty(G)) \to C^n(G, \mathbb{C})$ be defined by

$$(p\phi)(x_1, \ldots, x_n) = \phi(x_1, \ldots, x_n)(x_n^{-1} \ldots x_1^{-1})$$

for all $x_j \in G$ and all $\phi \in \mathcal{L}^n(\ell^1(G), \ell^\infty(G))$. Then p is a continuous linear map with $\|p\| = 1$. A direct calculation as above shows that, for all $\psi \in \mathcal{L}^{n-1}(\ell^1(G), \ell^\infty(G))$,

$$
\begin{aligned}
p\partial\psi(x_1, \ldots, x_n) &= \partial\psi(x_1, \ldots, x_n)((x_1 \ldots x_n)^{-1}) \\
&= \psi(x_2, \ldots, x_n)((x_2 \ldots x_n)^{-1}) \\
&\quad + \sum_{j=1}^{n-1} (-1)^j \psi(x_1, \ldots, x_j x_{j+1}, \ldots, x_n)((x_1 \ldots x_n)^{-1}) \\
&\quad + (-1)^n \psi(x_1, \ldots, x_{n-1})((x_1 \ldots x_{n-1})^{-1}) \\
&= p\psi(x_2, \ldots, x_n) \\
&\quad + \sum_{j=1}^{n-1} (-1)^j p\psi(x_1, \ldots, x_j x_{j+1}, \ldots, x_n) \\
&\quad + (-1)^n p\psi(x_1, \ldots, x_{n-1}) \\
&= \delta p\psi(x_1, \ldots, x_n)
\end{aligned}
$$

for all $x_1, \ldots, x_n$ in G. Further pP is the identity operator on $C_b^n(G)$, because

$$
\begin{aligned}
(pP\chi)(x_1, \ldots, x_n) &= P\chi(x_1, \ldots, x_n)((x_1 \ldots x_n)^{-1}) \\
&= \chi(x_1, \ldots, x_n),
\end{aligned}
$$

as $tr(1) = 1$, for all $x_1, \ldots, x_n$ in G and all $\chi \in C_b^n(G)$. This shows that $H_b^n(G, \mathbb{C})$ is embedded in $H^n(\ell^1(G), \ell^1(G))$.

(2) This proof is analogous to the one above. Direct calculations show that Q is a linear operator with $\|Q\| = 1$ and $\partial Q = Q\delta$. Define the map q from $\mathcal{L}^n(\ell^1(G), \ell^1(G))$ into $C^n(G, \mathbb{C})$ by

$$(q\phi)(x_1, \ldots, x_n) = tr(\psi(x_1, \ldots, x_n)((x_1 \ldots x_n)^{-1}))$$

for all $x_1, \ldots, x_n$ in G and ϕ in $\mathcal{L}^n(\ell^1(G), \ell^1(G))$. Then q is a continuous linear map with $\|q\| = 1$, qQ is the identity on $C_b^n(G)$, and $\delta q = q\partial$; the latter equality requires the tracial property of tr.

8.3.2 Problem. Exactly how are $H_b^n(G, \mathbb{C})$, and the two spaces $H^n(\ell^1(G), \ell^\infty(G))$ and $H^n(\ell^1(G), \ell^1(G))$ related? What other modules can be put in place of $\ell^1(G)$ and $\ell^\infty(G)$?

8.3.3 Corollary. If n is a positive integer greater than 1, then $H^2(\ell^1(\mathbf{F}_n),$ $\ell^1(\mathbf{F}_n)$ and $H^2(\ell^1(\mathbf{F}_n), \ell^\infty(\mathbf{F}_n))$ are non-zero.

Note that $H^1(\ell^1(G), \ell^1(G)) = 0$ for all discrete groups G by [JR] and that $H^1(\ell^1(G), \ell^\infty(G)) = 0$ by [J3]. Further observe that both of the modules $\ell^1(G)$ and $\ell^\infty(G)$ are dual modules.

8.3.4 Problem. Let G be a discrete group which could be assumed to be icc, where icc stands for infinite conjugacy class (i.e. each conjugacy class other than the identity is infinite). This condition on a group is equivalent to the group von Neumann algebra $VN(G)$ being a type II_1 factor; with the type II_1 following from the discreteness of the group.

How is $H_b^n(G)$ related to the Hochschild cohomology of $C_\lambda^*(G)$, the C^*-algebra of the left regular representation of G or $C^*(G)$, the full group C^*-algebra, or $VN(G)$, the von Neumann algebra generated by the left regular representation of G? In each case the cohomology is to be calculated with respect to suitable modules.

Observe that Theorem 8.3.1 is not valid if $\ell^1(G)$ is replaced by $VN(G)$ and $\ell^\infty(G)$ is replaced by $A(G) \cong VN(G)_*$ in (1) or $\ell^1(G)$ is replaced by $VN(G)$. The explanation is the following. Let G_1 be any discrete group, let G_0 be an amenable discrete icc group, and let $G = G_1 \times G_0$. Then

$$VN(G) = VN(G_1) \overline{\otimes} VN(G_0)$$

by [CowHa], [deCHa]. Further by [Co2, Prop. 6.8] the von Neumann algebra $VN(G_0)$ is injective and hence isomorphic to the hyperfinite type II_1 factor $\mathcal{R}$. Hence $H^n(VN(G), VN(G)) = 0$ by Theorem 6.2.3.

8.4 Problem List

The following is a list of problems concerning the Hochschild cohomology of a dual normal module over a von Neumann algebra. There is a vague grouping in the order of the problems but deep meanings should not be read into it. Numbers in brackets refer to the relevant section or result in these notes.

8.4.1. If $\mathcal{M}$ is a von Neumann algebra is $H^n(\mathcal{M}, \mathcal{M}) = 0$ for all $n \in \mathbb{N}$? This was conjectured by Kadison and Ringrose [KR2], [KR3]. It is known to be true except for certain classes of type II_1 von Neumann algebras. There is no information currently available for type II_1 factors $\mathcal{M}$ that cannot be written as $\mathcal{M} = \mathcal{M}_1 \overline{\otimes} \mathcal{M}_2$ with $\mathcal{M}_1$ and $\mathcal{M}_2$ type II_1 factors.

Is the answer yes for $n = 2$ or $n = 3$, where the methods arising from the non-commutative Grothendieck inequality of Haagerup and Pisier are available? (6.2, 6.3, 6.4.)

8.4.2. This problem runs in the opposite direction to 8.4.1. Is there a (finitely generated) countable discrete group G such that $H^2(VN(G), VN(G))$ is not zero? (8.1, 8.2.) Note that if the answer is yes there would be a type II_1 von Neumann algebra with no Cartan subalgebra (6.3).

8.4.3. If F_k denotes the free group on k (≥ 2) generators, is $H^2(VN(\mathsf{F}_k), VN(\mathsf{F}_k))$ equal to zero?

8.4.4. If $\mathcal{M}$ is a type II_1 factor with separable predual and if

$$\mathcal{M} \cong \mathcal{M} \otimes \mathsf{M}_k(\mathbb{C})$$

for some integer $k \geq 2$, is

$$H_w^n(\mathcal{M}, \mathcal{X}) \cong H_{wcb}^n(\mathcal{M}, \mathcal{X})$$

for all ultraweakly closed subspaces $\mathcal{X}$ of $B(H)$ satisfying $\mathcal{M} \cdot \mathcal{X} = \mathcal{X}$ and $\mathcal{X} \cdot \mathcal{M} = \mathcal{X}$? Here $H = L^2(\mathcal{M})$. If the fundamental group of a type II_1 factor is not trivial are the continuous and completely bounded cohomologies equal for dual normal completely bounded modules? (6.2) [ChES].

8.4.5. If $H^2(\mathcal{M}, \mathcal{X}) = 0$ for all dual normal modules, is $\mathcal{M}$ an injective von Neumann algebra? [Co1], [ChS6].

8.4.6. What cohomological properties does a von Neumann algebra $\mathcal{M}$ have that has finite Haagerup constant $\Lambda(\mathcal{M})$? What is the relation between the weak amenability condition of Cowling and Haagerup [CowHa] and Hochschild cohomology for C^*-algebras, von Neumann algebras or the various group algebras in the case of a countable discrete group. (Beware this definition of weak amenability is deep, depending on an approximation property with completely bounded operators and is not the straightforward weak amenability of Banach algebras, which is just $H^1(\mathcal{A}, \mathcal{A}^*) = 0$.)

8.4.7. The K-theory of a C^*-algebra is the study of the stabilized projections and invertibles in the algebra, where the stabilization is a limit over tensor products by $\mathsf{M}_n(\mathbb{C})$. Completely bounded operators are a stabilized operator theory by the same tensor product. Is there a connection between the K-theory for C^*-algebras, and completely bounded operators or completely bounded Hochschild cohomology?

8.4.8. Is there a relationship between property T for a von Neumann algebra and Hochschild cohomology? [CoJ].

8.4.9. Let $\mathcal{M}$ be a von Neumann algebra with separable predual. For which dual normal $\mathcal{M}$-modules $\mathcal{X}$ is $H_w^n(\mathcal{M}, \mathcal{X})$ a Banach space? The same question applies to completely bounded dual modules of the type mentioned in 8.4.4. Note that there is a separable C^*-algebra $\mathcal{A}$ such that $H^1(\mathcal{A}, \mathcal{A})$ is not a normed space, i.e. $\partial \mathcal{A}$ is not closed in $\mathcal{L}^1(\mathcal{A}, \mathcal{A})$. [KLR, Ex. 6.3].

8.4.10. Can we introduce a dual normal module over a von Neumann algebra $\mathcal{M}$ such that $H^1(\mathcal{M}, \mathcal{W}) = 0$ enables one to construct directly an increasing sequence of finite dimensional $*$-subalgebras in $\mathcal{M}$ whose union is weakly dense in $\mathcal{M}$? The point is to avoid having to pass through completely positive approximations and the existing theory.

8.4.11. If $\mathcal{N}$ is a von Neumann subalgebra of a von Neumann algebra $\mathcal{M}$, are $H^n(\mathcal{N}, \mathcal{M})$ and $H^n_{cb}(\mathcal{N}, \mathcal{M})$ zero? (4.3 and 6.2.)

8.4.12. If $\mathcal{M}$ is a von Neumann subalgebra of $B(H)$, what are

$$H^n(\mathcal{M}, B(H)/\mathcal{M}) \quad \text{and} \quad H^n_{cb}(\mathcal{M}, B(H)/\mathcal{M}),$$

particularly when $n = 1$?

8.4.13. If $\mathcal{N}$ is a von Neumann subalgebra of a von Neumann algebra $\mathcal{M}$ do derivations from $\mathcal{N}$ into $\mathcal{M}/\mathcal{N}$ lift to derivations from $\mathcal{N}$ into $\mathcal{M}$? The important case is $\mathcal{M} = B(H)$.

8.4.14. If a von Neumann algebra $\mathcal{M}$ has a Cartan subalgebra, is $H^n(\mathcal{M}, \mathcal{M})$ $= 0$ or $H^n(\mathcal{M}, B(H)) = 0$ for all n? (6.3.)

8.4.15. Let $\mathcal{M}$ be a type II_1 factor with injective subfactor $\mathcal{N}$ such that the relative commutant $\mathcal{N}' \cap \mathcal{M} = \mathbb{C}$. Let $\phi \colon \mathcal{M}^n \to \mathcal{M}_*$ be normal with $\partial \phi = 0$. Is there $\psi \colon \mathcal{M}^{n-1} \to \mathcal{M}_*$ such that

$$|(\phi - \partial \psi)(x_1, \ldots, x_n)(xy)| \le K \|x_1\| \ldots \|x_n\| \cdot \|x\|_2 \|y\|_2$$

for all $x_1, \ldots, x_n, x, y$?

Bibliography

[Ak] C.A. Akemann, The dual space of an operator algebra, *Trans. A.M.S.*, **126** (1967), 286–302.

[Al] E.M. Alfsen, *Compact convex sets and boundary integrals*, Springer Verlag, Berlin, 1971.

[Ar] W.B. Arveson, Subalgebras of C^*-algebras I, *Acta Math.*, **123** (1969), 141–224.

[B1] D.P. Blecher, Geometry of the tensor product of C^*-algebras, *Math. Proc. Cam. Phil. Soc.*, **104** (1988), 119–127.

[B2] D.P. Blecher, The standard dual of an operator space, *Pacific J. Math.*, **153** (1992), 15-30.

[B3] D.P. Blecher, Tensor products of operator spaces II, *Canad. J. Math.*, **44** (1992), 75-90.

[BP] D.P. Blecher and V.I. Paulsen, Tensor products of operator spaces, *J. Funct. Anal.*, **99** (1991), 262–292.

[BRS] D.P. Blecher, Z.J. Ruan and A.M. Sinclair, A characterization of operator algebras, *J. Funct. Anal.*, **89** (1990), 188–201.

[BS] D.P. Blecher and R.R. Smith, The dual of the Haagerup tensor product, *J. London Math. Soc.*, **45** (1992), 126–144.

[BD1] F.F. Bonsall and J. Duncan, *Numerical ranges of operators on normed spaces and of elements of normed algebras*, London Math. Soc. Lecture Note Series 2, Cambridge University Press, 1971.

[BD2] F.F. Bonsall and J. Duncan, *Complete normed algebras*, Springer Verlag, Berlin, 1973.

[Br] R. Brooks, Some remarks on bounded cohomology, *Ann. of Math. Studies*, **97** (1980), 53–63.

[Bu1] J.W. Bunce, Characterizations of amenable and strongly amenable C^*-algebras, *Pacific J. Math.*, **43** (1972), 563–572.

[Bu2] J.W. Bunce, Finite operators and amenable C^*-algebras, *Proc. A.M.S.*, **56** (1976), 145–151.

[Bu3] J.W. Bunce, The similarity problem for representations of C^*-algebras, *Proc. A.M.S.*, **81** (1981), 409–414.

[BuP1] J.W. Bunce and W.L. Paschke, Quasi-expectations and amenable von Neumann algebras, *Proc. A.M.S.*, **71** (1978), 232–236.

[BuP2] J.W. Bunce and W.L. Paschke, Derivations on a C^*-algebra and its double dual, *J. Funct. Anal.*, **37** (1980), 235–247.

[CE1] M.D. Choi and E.G. Effros, Injectivity and operator spaces, *J. Funct. Anal.*, **24** (1977), 156–209.

[CE2] M.D. Choi and E.G. Effros, Lifting problems and the cohomology of C^*-algebras, *Canad. J. Math.*, **29** (1977), 1092–1111.

[Ch1] E. Christensen, Perturbations of operator algebras, *Invent. Math.*, **43** (1977), 1–13.

[Ch2] E. Christensen, Perturbations of operator algebras II, *Indiana Math. J.*, **26** (1977), 891–904.

[Ch3] E. Christensen, Extensions of derivations, *J. Funct. Anal.*, **27** (1978), 234–247.

[Ch4] E. Christensen, On non-self-adjoint representations of C^*-algebras, *Amer. J. Math.*, **103** (1981), 817–833.

[Ch5] E. Christensen, Extensions of derivations II, *Math. Scand.*, **50** (1982), 111–122.

[Ch6] E. Christensen, Derivations and their relation to perturbations of operator algebras, *Proc. Symp. Pure Math.*, **38** (1982), 261–273.

[Ch7] E. Christensen, Similarities of II_1 factors with property Γ, *J. Operator Theory*, **15** (1986), 281–288.

[ChES] E. Christensen, E.G. Effros and A.M. Sinclair, Completely bounded multilinear maps and C^*-algebraic cohomology, *Invent. Math.*, **90** (1987), 279–296.

[ChEv] E. Christensen and D.E. Evans, Cohomology of operator algebras and quantum dynamical semigroups, *J. London Math. Soc.*, **20** (1979), 358–368.

[ChPSS] E. Christensen, F. Pop, A.M. Sinclair and R.R. Smith, On the cohomology groups of certain finite von Neumann algebras, preprint.

[ChS1] E. Christensen and A.M. Sinclair, Representations of completely bounded multilinear operators, *J. Funct. Anal.*, **72** (1987), 151–181.

[ChS2] E. Christensen and A.M. Sinclair, A survey of completely bounded operators, *Bull. London Math. Soc.*, **21** (1989), 417–448.

[ChS3] E. Christensen and A.M. Sinclair, On the vanishing of $H^n(A, A^*)$ for certain C^*-algebras, *Pacific J. Math.*, **137** (1989), 55–63.

[ChS4] E. Christensen and A.M. Sinclair, On the Hochschild cohomology for von Neumann algebras, preprint.

[ChS5] E. Christensen and A.M. Sinclair, Module mappings into von Neumann algebras and injectivity, preprint.

[ChS6] E. Christensen and A.M. Sinclair, A cohomological characterisation of approximately finite dimensional von Neumann algebras, preprint.

[ChS7] E. Christensen and A.M. Sinclair, On von Neumann algebras which are complemented subspaces of $B(H)$, *J. Funct. Anal.*, **122** (1994), 91–102.

[Co1] A. Connes, On the cohomology of operator algebras, *J. Funct. Anal.*, **28** (1978), 248–253.

[Co2] A. Connes, Classification of injective factors, *Ann. Math.*, **104** (1976), 73–115.

[Co3]   A. Connes, Non-commutative differential geometry, *Publ. I.H.E.S.*,
        62 (1985), 41–144.

[CoJ]   A. Connes and V. Jones, Property T for von Neumann algebras,
        Bull. London Math. Soc., **17** (1985), 57–62.

[CowHa] M. Cowling and U. Haagerup, Completely bounded multipliers of
        the Fourier algebra of a simple Lie group of real rank one, *Invent.
        Math.*, **96** (1989), 507–549.

[Cr1]   I.G. Craw, Axiomatic cohomology for Banach modules, *Proc.
        A.M.S.*, **38** (1973), 68–74.

[Cr2]   I.G. Craw, Axiomatic cohomology of operator algebras, *Bull. Soc.
        Math. France*, **101** (1973), 449–460.

[deCHa] J. de Cannière and U. Haagerup, Multipliers of the Fourier al-
        gebras of some simple Lie groups and their discrete subgroups,
        Amer. J. Math., **107** (1984), 455–500.

[Di1]   J. Dixmier, Quelques propriétés de suites centrales dans les fac-
        teurs de type II_1, *Invent. Math.*, **7** (1969), 215–225.

[Di2]   J. Dixmier, *Les algèbres d'opérateurs dans l'espace Hilbertien*,
        Gauthier-Villars, Paris, 1969.

[Di3]   J. Dixmier, C^*-*algebras*, North Holland, New York, 1977.

[E1]   E.G. Effros, Property Γ and inner amenability, *Proc. A.M.S.*, **47**
        (1975), 483–486.

[E2]   E.G. Effros, Advances in quantized functional analysis, *Proceed-
        ings of the International Congress of Mathematicians*, Berkeley,
        1986.

[E3]   E.G. Effros, On multilinear completely bounded maps, *Contemp.
        Math.*, **62** (1987), 479–501.

[E4]   E.G. Effros, Amenability and virtual diagonals for von Neumann
        algebras, *J. Funct. Anal.*, **78** (1988), 137–153.

[EK]   E.G. Effros and A. Kishimoto, Module maps and Hochschild-
        Johnson cohomology, *Indiana Math. J.*, **36** (1987), 257–276.

[EL]   E.G. Effros and E.C. Lance, Tensor products of operator algebras,
        Adv. in Math., **25** (1977), 1–34.

[ER1]   E.G. Effros and Z.J. Ruan, On matrically normed spaces, *Pacific
        J. Math.*, **132** (1988), 243–264.

[ER2]   E.G. Effros and Z.J. Ruan, A new approach to operator spaces,
        Canad. Math. Bull., **34** (1991), 329–337.

[ER3]   E.G. Effros and Z.J. Ruan, On approximation properties of oper-
        ator spaces, *Internat. J. Math.*, **1** (1990), 163–187.

[ER4]   E.G. Effros and Z.J. Ruan, Recent developments in operator spaces,
        Proceedings of the Satellite Conference of ICM-90, World Scien-
        tific, Singapore, 1991, pp. 146–164.

[ER5] E.G. Effros and Z.J. Ruan, *Operator Spaces*, book manuscript in preparation.

[El1] G.A. Elliott, On the classification of inductive limits of sequences of semi-simple finite dimensional algebras, *J. Algebra*, **38** (1976), 29–44.

[El2] G.A. Elliott, On approximately finite dimensional von Neumann algebras II, *Canad. Math. Bull.*, **21** (1978), 415-418.

[El3] G.A. Elliott, On derivations of AW^*-algebras, *Tôhoku Math. J.*, **30** (1978), 263–276.

[Fa] V.A. Faiziev, Pseudocharacters on a free group and some group constructions, *Russian Math. Surveys*, **43** (1988), 225–226.

[FM] J. Feldman and C.C. Moore, Ergodic equivalence relations, cohomology and von Neumann algebras, *Trans. A.M.S.*, **234** (1977), 289–361.

[Ga] P. Gajendragadkar, Ph.D. Thesis, Indiana University, 1970.

[GH] M. Gerstenhaber and M. Hazewinkel, Deformation theory of algebras and structures and applications, *NATO ASI Series Math. and Phys.*, **247** (1988).

[GHL] F.L. Gilfeather, A. Hopenwasser and D.R. Larson, Reflexive algebras with finite width lattices: tensor products, cohomology, compact perturbations, *J. Funct. Anal.*, **55** (1984), 176–199.

[GS1] F.L. Gilfeather and R.R. Smith, Operator algebras with arbitrary Hochschild cohomology, *Contemp. Math.*, **120** (1991), 33–40.

[GS2] F.L. Gilfeather and R.R. Smith, Cohomology for operator algebras: cones and suspensions, *Proc. London Math. Soc.*, **65** (1992), 175–198.

[GS3] F.L. Gilfeather and R.R. Smith, Cohomology for operator algebras, *Proceedings of the International Workshop on Elementary Operators and Applications*, 189–195, World Scientific, Singapore, 1992.

[GS4] F.L. Gilfeather and R.R. Smith, Cohomology for operator algebras: joins, *Amer. J. Math.*, **116** (1994), 541–561.

[Gl] J. Glimm, A Stone-Weierstrass theorem for C^*-algebras, *Ann. Math.*, **72** (1960), 216–244.

[Gre] F. Greenleaf, *Invariant means on topological groups*, Van Nostrand, New York, 1969.

[Gri1] R.I. Grigorchuk, Some remarks on bounded cohomology, Talk to ICMS Conference, Edinburgh, April 1993.

[Gri2] R.I. Grigorchuk, Some results on bounded cohomology, preprint.

[Grom] M. Gromov, Volume and bounded cohomology, *Publ. Math. IHES*, **56** (1982), 5–100.

[Grot] A. Grothendieck, Résumé de la théorie métrique des produits ten-
        soriels topologiques, *Bol. Soc. Math. São Paulo*, **8** (1956), 1–79.

[Gu]   A. Guichardet, Sur l'homologie et la cohomologie des algèbres de
        Banach, *C.R. Acad. Sci. Paris*, **262** (1966), 38–41.

[Ha1]   U. Haagerup, All nuclear C^*-algebras are amenable, *Invent. Math.*,
        74 (1983), 305–319.

[Ha2]   U. Haagerup, Solution of the similarity problem for cyclic repre-
        sentations of C^*-algebras, *Ann. Math.*, **118** (1983), 215–240.

[Ha3]   U. Haagerup, The Grothendieck inequality for bilinear forms on
        C^*-algebras, *Adv. in Math*, **56** (1985), 93–116.

[Ha4]   U. Haagerup, A new proof of the equivalence of injectivity and
        hyperfiniteness for factors on a separable Hilbert space, *J. Funct.
        Anal.*, **62** (1985), 160–201.

[Ha5]   U. Haagerup, Injectivity and decomposition of completely bounded
        maps, Springer Lecture Notes in Math., **1132**, Springer-Verlag,
        1985, pp. 170–222.

[Ha6]   U. Haagerup, The α-tensor product for C^*-algebras, unpublished
        manuscript.

[Ha7]   U. Haagerup, Decomposition of completely bounded maps on op-
        erator algebras, unpublished manuscript.

[Hal]   H. Halpern. Irreducible module homomorphisms of a von Neu-
        mann algebra into its center, *Trans. A.M.S.*, **140** (1969), 195–221.

[He]   A. Helemskii, *The homology of Banach and topological algebras*,
        Kluwer, Dordrecht, 1989.

[HewR] E. Hewitt and K.A. Ross, *Abstract harmonic analysis*, Vol. I,
        Springer Verlag, Berlin, 1963.

[HiT]   M.W. Hirsch and W.P. Thurston, Foliated bundles, invariant mea-
        sures and flat manifolds, *Math. Ann.*, **101** (1975), 777–781.

[Ho1]   G. Hochschild, On the cohomology groups of an associative alge-
        bra, *Ann. Math.*, **46** (1945), 58–67.

[Ho2]   G. Hochschild, On the cohomology theory for associative algebras,
        Ann. Math., **47** (1946), 568–579.

[Ho3]   G. Hochschild, Cohomology and representations of associative al-
        gebras, *Duke Math. J.*, **14** (1947), 921–948.

[Iv1]   N.V. Ivanov, Foundation of the theory of bounded cohomology, *J.
        Soviet Math.*, **143** (1985), 69–109.

[Iv2]   N.V. Ivanov, Second bounded cohomology group, *J. Soviet Math.*,
        167 (1988), 117–120.

[J1]   B.E. Johnson, The Wedderburn decomposition of Banach algebras
        with finite dimensional radical, *Amer. J. Math.*, **90** (1968), 866–
        876.

[J2] B.E. Johnson, Continuity of derivations on commutative Banach algebras, *Amer. J. Math.*, **91** (1969), 1–10.

[J3] B.E. Johnson, Cohomology in Banach algebras, *Memoirs A.M.S.*, **127** (1972), pp. 96.

[J4] B.E. Johnson, Approximate diagonals and cohomology of certain annihilator Banach algebras, *Amer. J. Math.*, **94** (1972), 685–698.

[J5] B.E. Johnson, A class of II_1 factors without property P but with zero second cohomology, *Arkiv för Math.*, **12** (1974), 153–159.

[J6] B.E. Johnson, Perturbations of Banach algebras, *Proc. London Math. Soc.*, **34** (1977), 439–458.

[J7] B.E. Johnson, Weak amenability of group algebras, *Bull. London Math. Soc.*, **23** (1991), 281–284.

[JKR] B.E. Johnson, R.V. Kadison and J.R. Ringrose, Cohomology of operator algebras III. Reduction to normal cohomology, *Bull. Soc. Math. France*, **100** (1972), 73–96.

[JP] B.E. Johnson and S.K. Parrott, Operators commuting modulo the set of compact operators with a von Neumann algebra, *J. Funct. Anal.*, **11** (1972), 39–61.

[JR] B.E. Johnson and J.R. Ringrose, Derivations of operator algebras and discrete group algebras, *Bull. London Math. Soc.*, **1** (1969), 70–74.

[JS] B.E. Johnson and A.M. Sinclair, Continuity of derivations and a problem of Kaplansky, *Amer. J. Math.*, **90** (1968), 1067–1073.

[K] L. Kadison, A relative cyclic cohomology theory useful for computations, *C.R. Acad. Sci. Paris*, Serie I, **308** (1989), 569–573.

[Ka1] R.V. Kadison, Unitary invariants for representations of operator algebras, *Ann. Math.*, **66** (1957), 304–379.

[Ka2] R.V. Kadison, Derivations of operator algebras, *Ann. Math.*, **83** (1966), 280–293.

[Ka3] R.V. Kadison, A note on derivations of operator algebras, *Bull. London Math. Soc.*, **7** (1975), 41-44.

[Ka4] R.V. Kadison, On an inequality of Haagerup–Pisier, *J. of Operator Theory*, **29** (1993), 57–67.

[KK] R.V. Kadison and D. Kastler, Perturbations of von Neumann algebras I, stability of type, *Amer. J. Math.*, **94** (1972), 38–54.

[KLR] R.V. Kadison, E.C. Lance and J.R. Ringrose, Derivations and automorphisms of operator algebras II, *J. Funct. Anal.*, **1** (1967), 204–221.

[KR1] R.V. Kadison and J.R. Ringrose, Derivations and automorphisms of operator algebras, *Comm. Math. Phys.*, **4** (1967), 32–63.

[KR2] R.V. Kadison and J.R. Ringrose, Cohomology of operator algebras I. Type I von Neumann algebras, *Acta Math.*, **126** (1971), 227–

243.

[KR3] R.V. Kadison and J.R. Ringrose, Cohomology of operator algebras II. Extended cobounding and the hyperfinite case, *Arkiv för Math.*, **9** (1971), 55–63.

[KR4] R.V. Kadison and J.R. Ringrose, *Fundamentals of the theory of operator algebras*, Vols. I and II, Academic Press, New York, 1983.

[KS] S. Kaijser and A.M. Sinclair, Projective tensor products of C^*-algebras, *Math. Scand.*, **55** (1984), 161–187.

[Kam] H. Kamowitz, Cohomology groups of commutative Banach algebras, *Trans. A.M.S.*, **102** (1962), 352–372.

[Kap] I. Kaplansky, Modules over operator algebras, *Amer. J. Math.*, **75** (1953), 839–859.

[KrS] J. Kraus and S.D. Schack, The cohomology and deformations of CSL-algebras: a précis (summary of forthcoming manuscript).

[L] E.C. Lance, Cohomology and perturbations of nest algebras, *Proc. London Math. Soc.*, **43** (1981), 334–356.

[Li] G. Lindblad, Dissipative operators and cohomology of operator algebras, *Lett. Math. Phys.*, **1** (1976), 219–224.

[McD] D. McDuff, Central sequences and the hyperfinite factor, *Proc. London Math. Soc.*, **21** (1970), 443–461.

[MvN1] F.J. Murray and J. von Neumann, On rings of operators, *Ann. of Math.*, **37** (1936), 116–229.

[MvN2] F.J. Murray and J. von Neumann, On rings of operators IV, *Ann. Math.*, **44** (1943), 716–808.

[N] J.P. Nielsen, Cohomology of some non-self-adjoint operator algebras, *Math. Scand.*, **47** (1980), 150–156.

[P1] A.L.T. Paterson, *Amenability*, Mathematical Surveys and Monographs, No. 29, A.M.S., Providence, 1988.

[P2] A.L.T. Paterson, Invariant mean characterizations of amenable operator algebras, preprint.

[Pa1] V.I. Paulsen, Every completely polynomially bounded operator is similar to a contraction, *J. Funct. Anal.*, **55** (1984), 1–17.

[Pa2] V.I. Paulsen, *Completely bounded maps and dilations*, Notes in Mathematics Series 146, Pitman, New York, 1986.

[PS] V.I. Paulsen and R.R. Smith, Multilinear maps and tensor norms on operator systems, *J. Funct. Anal.*, **73** (1987), 258–276.

[Pe1] G.K. Pedersen, Lifting derivations from quotients of separable C^*-algebras, *Proc. Nat. Acad. Sci. U.S.A.*, **73** (1976), 1414–1415.

[Pe2] G.K. Pedersen, C^*-*algebras and their automorphism groups*, Academic Press, New York, 1979.

[PhR1] J. Phillips and I. Raeburn, Perturbations of C^*-algebras II, *Proc. London Math. Soc.*, **43** (1981), 46–72.

[PhR2] J. Phillips and I. Raeburn, Central cohomology of C^*-algebras, *J. London Math. Soc.*, **28** (1983), 363–375.

[Pie] J.P. Pier, *Amenable Banach algebras*, Pitman Research Notes in Math., No. 172, New York, 1988.

[Pi1] G. Pisier, Grothendieck's theorem for non-commutative C^*-algebras with an appendix on Grothendieck's constant, *J. Funct. Anal.*, **29** (1978), 397–415.

[Pi2] G. Pisier, *Factorization of linear operators and the geometry of Banach spaces*, CBMS Series No. 60, A.M.S., Providence, R.I., 1986.

[Pi3] G. Pisier, The operator Hilbert space OH, complex interpolation and tensor norms, preprint.

[PoS] F. Pop and R.R. Smith, Cohomology for certain finite factors, *Bull. London Math. Soc.*, to appear.

[Pop1] S. Popa, On a problem of R.V. Kadison on maximal abelian *-subalgebras in factors, *Invent. Math.*, **65** (1981), 269–281.

[Pop2] S. Popa, Notes on Cartan subalgebras in type II_1 factors, *Math. Scand.*, **57** (1985), 171–188.

[Pop3] S. Popa, The commutant modulo the set of compact operators of a von Neumann algebra, *J. Funct. Anal.*, **71** (1987), 393–408.

[PopR] S. Popa and F. Rădulescu, Derivations on von Neumann algebras into the compact ideal space of a semifinite algebra, *Duke Math. J.*, **57** (1988), 485–518.

[R1] F. Rădulescu, Vanishing of $H_w^2(M, K(H))$ for certain finite von Neumann algebras, *Trans. Amer. Math. Soc.*, **326** (1991), 569–584.

[R2] F. Rădulescu, Singularity of the radial subalgebra of $\mathcal{L}(F_N)$ and the Pukansky invariant, *Pacific J. Math.*, **151** (1991), 297–306.

[RaT] I. Raeburn and J.L. Taylor, Hochschild cohomology and perturbations of Banach algebras, *J. Funct. Anal.*, **25** (1977), 258–266.

[Ri1] J.R. Ringrose, Automatic continuity of derivations of operator algebras, *J. London Math. Soc.*, **5** (1972), 432–438.

[Ri2] J.R. Ringrose, Lectures on the trace in a finite von Neumann algebra, *Lecture Notes in Mathematics*, **247**, 309–354, Springer-Verlag, Berlin, 1972.

[Ri3] J.R. Ringrose, Cohomology of operator algebras, *Lecture Notes in Mathematics* **247**, 355–433, Springer Verlag, Berlin, 1972.

[Ri4] J.R. Ringrose, Linear mappings between operator algebras, *Symp. Math.*, **20**, 297–315, Academic Press, London, 1976.

[Ri5] J.R. Ringrose, Derivations of quotients of von Neumann algebras, *Proc. London Math. Soc.*, **36** (1978), 1–26.

[Ri6] J.R. Ringrose, Cohomology theory for operator algebras, *Proc. Symp. Pure Math.*, **38** (1982), 229–252.

[Ru] Z.J. Ruan, Subspaces of C^*-algebras, *J. Funct. Anal.*, **76** (1988), 217–230.

[S1] S. Sakai, Derivations of W^*-algebras, *Ann. Math.*, **83** (1966), 273–279.

[S2] S. Sakai, *C^*-algebras and W^*-algebras*, Springer Verlag, Berlin, 1971.

[S3] S. Sakai, *Operator algebras in dynamical systems*, Cambridge University Press, 1991.

[Si1] A.M. Sinclair, Annihilator ideals in the cohomology of Banach algebras, *Proc. A.M.S.*, **33** (1972), 361–366.

[SiSm] A.M. Sinclair and R.R. Smith, Cartan subalgebras of finite von Neumann algebras, preprint.

[Sm1] R.R. Smith, Completely bounded multilinear maps and Grothendieck's inequality, *Bull. London Math. Soc.*, **20** (1988), 606–612.

[Sm2] R.R. Smith, Completely bounded module maps and the Haagerup tensor product, *J. Funct. Anal.*, **102** (1991), 156–175.

[SmW] R.R. Smith and J.D. Ward, Matrix ranges for Hilbert space operators, *Amer. J. Math.*, **102** (1980), 1031-1081.

[St] J. Stampfli, The norm of a derivation, *Pacific J. Math.*, **33** (1970), 737–747.

[Sti] W.F. Stinespring, Positive functions on C^*-algebras, *Proc. A.M.S.*, **6** (1955), 211–216.

[Str] S. Strătilă, *Modular theory in operator algebras*, Abacus Press, Tunbridge Wells, 1981.

[StrZ] S. Strătilă and L. Zsidó, *Lectures on von Neumann algebras*, Abacus Press, Tunbridge Wells, 1979.

[T] M. Takesaki, *Theory of operator algebras I*, Springer Verlag, Berlin, 1979.

[Ta] J.L. Taylor, Homology and cohomology for topological algebras, *Adv. in Math.*, **9** (1972), 137–182.

[Th] A.M. Thorpe, Nuclear C^*-algebras and injective von Neumann algebras, unpublished notes.

[To] J. Tomiyama, On the projections of norm one in W^*-algebras I, II, III, *Proc. Japan Acad.*, **33** (1957) 608–612; *Tôhoku Math. J.*, **10** (1958) 204–209; *Tôhoku Math. J.*, **11** (1959), 125–129.

[V] A.M. Vershik, Non-measurable decompositions, orbit theory, algebras of operators, *Dokl. Akad. Nauk S.S.S.R.*, **199** (1971), 1004–1007.

[W1] G. Wittstock, Ein operatorwertiger Hahn-Banach Satz, *J. Funct. Anal.*, **40** (1981), 127–150.

[W2] G. Wittstock, Extensions of completely bounded C^*-module ho-
      momorphisms, in *Proc. Conference on Operator Algebras and Group
      Representations*, Neptun 1980, Pitman, New York, 1983.

[W3] G. Wittstock, On matrix order and convexity, Functional Analy-
      sis, surveys and recent results, *Math. Studies*, **90** (1984), 175–188.

[Wo] M. Wodzicki, Vanishing of cyclic homology of stable C^*-algebras,
      C.R. Acad. Sci. Paris, **307** (1988), 329–334.

[ZeM] G. Zeller-Meier, Sur les automorphismes des algèbres de Banach,
      C.R. Acad. Sci. Paris Sér A-B, **264** (1967), 1131–1132.

[Zs] L. Zsidó, The norm of a derivation in a W^*-algebra, *Proc. A.M.S.*,
      38 (1973), 147–150.

Notation

Numbers indicate the pages on which the definitions can be found.

$\mathbb{N}, \mathbb{Z}, \mathbb{Q}, \mathbb{R}, \mathbb{C}$	denote respectively the positive integers, integers, rationals, real numbers, complex numbers.
$H, K, \ldots$	denote Hilbert spaces with inner product $\langle \cdot, \cdot \rangle$
$\xi, \eta, \ldots$	denote vectors in Hilbert spaces.
$\mathcal{M}, \mathcal{N}, \ldots$	denote von Neumann algebras.
$\mathcal{M}', \mathcal{N}', \ldots$	denote commutants of von Neumann algebras.
$\mathcal{R}$	denotes a hyperfinite von Neumann algebra.
$\mathcal{A}, \mathcal{B}, \ldots$	denote maximal abelian self-adjoint subalgebras (masas) of von Neumann algebras, and sometimes C^*-algebras in Chapter 3.
$\otimes$	denotes the algebraic tensor product.
$\bar{\otimes}, \otimes_{\min}, \otimes_{\max}, \otimes_h$	denote respectively the spatial von Neumann, the minimal, the maximal, and the Haagerup tensor products.
tr	denotes the trace on a finite von Neumann algebra.
$x, y, z, \ldots, a, b, c \ldots$	denote elements of algebras.
$B(H, K)$	denotes the space of bounded operators from H to K.
$\pi, \theta, \rho \ldots$	denote $*$-representations.
$\phi, \psi, \chi, \theta, \omega,$	denote maps between algebras or modules.
$\mathcal{V}, \mathcal{W}, \ldots$	denote Banach modules over operator algebras.
$\mathcal{L}^n(\mathcal{M}, \mathcal{V})$	denotes bounded n-linear maps from $\mathcal{M}^n$ to $\mathcal{V}$, 2, 9.
$\mathcal{L}_w^n(\mathcal{M}, \mathcal{V})$	denotes normal n-linear maps from $\mathcal{M}^n$ to a dual module $\mathcal{V}$, 3.
$\mathcal{L}_{cb}^n(\mathcal{M}, \mathcal{V})$	denotes completely bounded n-linear maps from $\mathcal{M}^n$ to $\mathcal{V}$.
$\mathcal{L}_{wcb}^n(\mathcal{M}, \mathcal{V})$	denotes completely bounded normal n-linear maps from $\mathcal{M}^n$ to $\mathcal{V}$.
$\mathcal{L}^n(\mathcal{M}, \mathcal{V}: /\mathcal{A})$	denotes those elements of $\mathcal{L}^n(\mathcal{M}, \mathcal{V})$ which are multimodular with respect to $\mathcal{A}$ with similar interpretations when subscripts "w", "cb" etc. are added to $\mathcal{L}^n$.
$\mathcal{L}_{cb}(\mathcal{A}, \mathcal{M})_{\mathcal{M}}$	denotes completely bounded linear maps which are right-modular with respect to $\mathcal{M}$.
$H^n(\mathcal{M}, \mathcal{V})$	denotes the nth Hochschild cohomology group of $\mathcal{M}$ with values in $\mathcal{V}$.

$H_w^n(\mathcal{M}, \mathcal{V}), H_{cb}^n(\mathcal{M}, \mathcal{V}),$ $H_{wcb}^n(\mathcal{M}, \mathcal{V})$	denote respectively the normal, completely bounded, and normal and completely bounded versions of $H^n(\mathcal{M}, \mathcal{V})$, 10.
$\mathbf{M}_n$ or $\mathbf{M}_n(\mathbf{C})$	denote the $n \times n$ scalar matrices, 12.
$\mathbf{M}_{m,n}(\mathcal{A})$	denotes the rectangular $m \times n$ matrices over an algebra $\mathcal{A}$, 29.
$\| \cdot \|_{cb}$	denotes the completely bounded norm, 12.
∂	denotes the Hochschild coboundary operator.
Ker ∂	denotes the kernel of ∂.
Im ∂	denotes the image of ∂.
δ	denotes the coboundary operator in group cohomology, 170.
$H_b^n(G)$	denotes the nth bounded cohomology group of G with values in $\mathbf{R}$, 171.
dist$(T, \mathcal{M})$	denotes the distance from an operator T to an algebra $\mathcal{M}$, 71.

Index

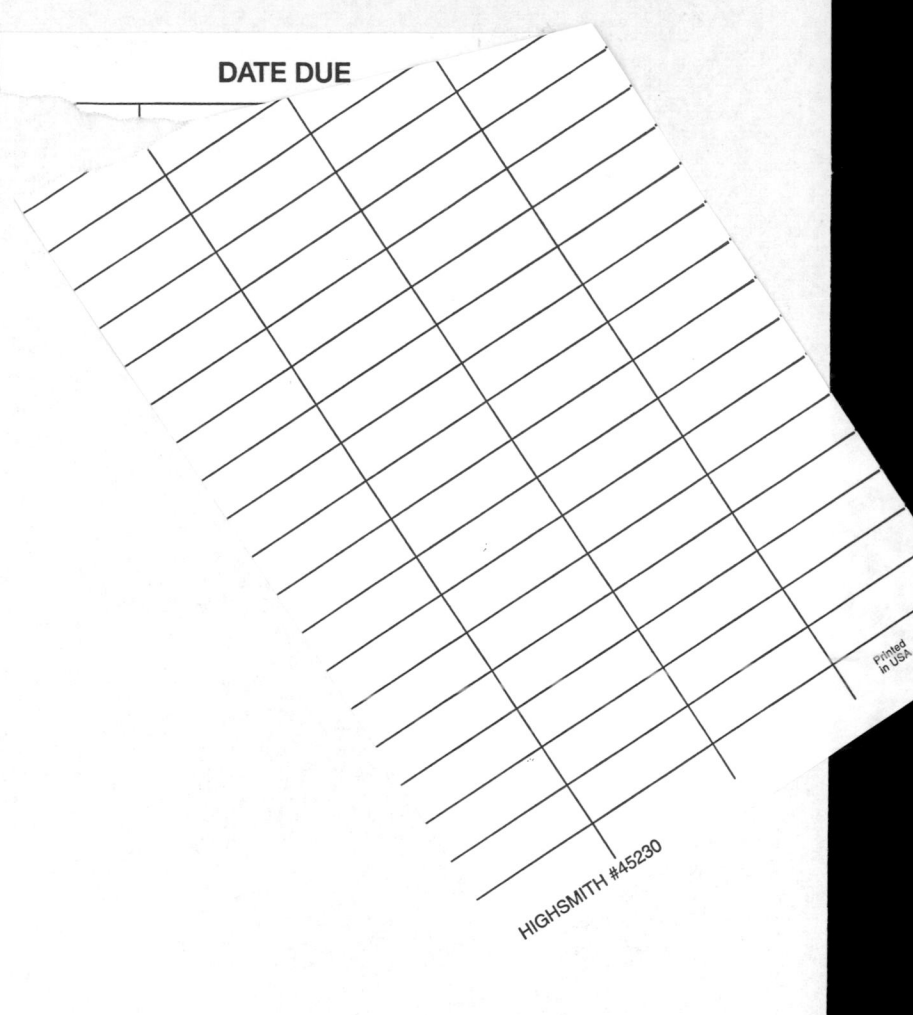

DATE DUE

HIGHSMITH #45230

Printed in USA